ON THE NATURE OF LIMBS

ON THE NATURE OF LIMBS

A Discourse

Richard Owen

WITH A PREFACE BY
Brian K. Hall

WITH INTRODUCTORY ESSAYS BY
*Ron Amundson, Kevin Padian,
Mary P. Winsor, and Jennifer Coggon*

THE UNIVERSITY OF CHICAGO PRESS
Chicago and London

Ron Amundson is professor of philosophy at the University of Hawaii at Hilo.

Kevin Padian is professor of integrative biology, curator in the Museum of Paleontology, and director of the College Writing Programs at the University of California, Berkeley.

Mary P. Winsor is emerita professor of zoology at the University of Toronto.

Jennifer Coggon is a student and researcher at the University of Toronto.

The University of Chicago Press, Chicago 60637
The University of Chicago Press, Ltd., London
© 2007 by The University of Chicago
All rights reserved. First published 1849
University of Chicago Press edition 2007
Printed in the United States of America

16 15 14 13 12 11 10 09 08 07   1 2 3 4 5

ISBN-13: 978-0-226-64194-2 (cloth)
ISBN-13: 978-0-226-64193-5 (paper)
ISBN-10: 0-226-64194-5 (cloth)
ISBN-10: 0-226-64193-7 (paper)

Library of Congress Cataloging-in-Publication Data

Owen, Richard, 1804–1892.
    On the nature of limbs : a discourse / Richard Owen with introductory essays by Ron Amundson . . . [et. al].
        p. cm.
    Includes bibliographical references and index.
    ISBN-13: 978-0-226-64194-2 (cloth : alk. paper)
    ISBN-13: 978-0-226-64193-5 (pbk. : alk. paper)
    ISBN-10: 0-226-64194-5 (cloth : alk. paper)
    ISBN-10: 0-226-64193-7 (pbk. : alk. paper)
    1. Extremities (Anatomy)—Evolution.   I. Title.
QL950.7O94 2007
571.3'1—dc22

2007009519

CONTENTS

PREFACE

Brian K. Hall

> But from the Epicurean slough of despond every healthy
> mind naturally recoils: and reverting therefore to the hypoth-
> esis of the dependence of special and serial homologies upon
> some wider principle of conformity, we have . . . to enquire,
> what is the archetype or essential nature of the limbs. (Owen
> 1849, 40)

"On The Nature of Limbs," a discourse delivered on Friday, February 9, 1849, at an evening meeting of the Royal Institution of Great Britain by Richard Owen, FRS, stands as a classic analysis of the nature of limbs and fins as homologous elements.

In a tradition that can be traced back to the writings of Aristotle, Owen was concerned with the essential nature of fins and limbs as homologous elements; Owen defined his lecture as "On the general and Serial Homologies of the Locomotive Extremities" (2).[1] Owen followed Aristotle, who saw that "birds in a way resemble fishes. For birds have their wings in the upper part of their bodies and fishes have two fins in the front part of their bodies. Birds have feet on their

1. Pagination is to the original page numbers in Owen (1849) as reproduced in this volume.

under part and most fishes have a second pair of fins in their under-part and near their front fins" (Aristotle, *De Incessu Animalium,* 724 BC, cited from Moore, 1993, 34).

Like Aristotle, Owen was concerned with the homologous relationships between fins and limbs, with their essential nature and unity of type: "The 'limbs' . . . are the parts called the 'arms' and 'legs' in Man; the 'fore-' and 'hind-legs' of beasts; the 'wings' and 'legs' of Bats and Birds; the 'pectoral fins' and 'ventral [pelvic] fins' of Fishes" (3). Owen took as given that "the arm of the Man is the fore-leg of the Beast, the wing of the Bird, and the pectoral fin of the Fish" (3) and that these are homologous parts, a relationship that fascinates us to this day.[2]

Owen used "Nature" in the title of his talk: "in the sense of the German *'Bedeutung'* [signification] as signifying that essential character of a part which belongs to it in its relation to a predetermined pattern, answering to the 'idea' of the Archetypal World in the Platonic cosmogony, which archetype or primal pattern is the basis supporting all the modifications of such part" (2–3). So, like Aristotle, Owen was concerned with transformation *within* a type. But, as discussed by Ron Amundson and Kevin Padian in their introductory essays, we also find glimmers in Owen's lecture of transformation *between* types: "'Nature' . . . has advanced with slow and stately steps, guided by the archetypal light" (86).

The transformation of fins to limbs comes so easily to our understanding these days that it takes some effort to realize that, in Owen's day, and despite their evident

2. See Hinchliffe and Johnson (1980), Hinchliffe et al. (1991), Hall (1994, 2005), Carroll (1997), Zimmer and Buell (1998), and Clack (2002) for analyses spanning the past quarter of a century.

differences, *fins and limbs were united by Aristotelian typology rather than as a transformation series*. The seeds of transformation contained in Owen's 1849 lecture, coupled with the depth and beauty of his analysis of fins and limbs, and the resurgence of interest in fins and limbs in developmental, evolutionary, and evolutionary developmental biology, all make this reissue of Owen's essay especially timely; several topics treated by Owen have become entire fields of investigation. Seven aspects of Owen's essay that I see as important are (with page numbers in parentheses):

1. The value of the essay as an exposition on the nature of limbs within and between types. So we find Owen speaking of the burrowing limb of the mole as "not very different in form and character from the fin" (6).

2. Owen takes us through theories created to explain the origin of the limbs that included liberated ribs (Oken), modified vertebrae (Carus), and modified gill arches or opercular bones (Geoffroy).

3. The essay provides an important pre-Darwinian analysis of the adaptation of "parts . . . serving chiefly for locomotion, but sometimes adapted to other offices" (4) and shows that Owen was concerned "that the principle of final adaptation fails to satisfy all the conditions of the problem" (39).

4. Owen provides an insight into the mid-nineteenth century view of endo- and exoskeletons, "the dermo- or exo-skeleton" being regarded as the "hardened skin . . . in the Invertebrata" (4).

5. Insights into Owen's approach to complexity are revealed to us: "The simplest form of the locomotive member is that of the fin" (5).

6. Owen saw that when digits are lost in evolution "the first or innermost digit, as a general rule, is the first to disappear . . . the outer digit v and v is the next to disappear" (35).

7. Owen's essay provides an important illustration of his use of the concepts of special, serial, and general homology. Owen clarifies his conception of

serial homology: "I . . . call the bones so related serially in the same skeleton 'homotypes,' and restrict
the term 'homologue' to the corresponding bones in
different individuals or species" (17; see also Owen
1848). Pectoral and pelvic extremities are serial
homologues (73ff.).

Delineating homology from analogy is one of Owen's
most remembered contributions. So too was his recognition of the homology of fins and limbs. Today we
recognize the homology of fins and limbs at all the
major levels in the biological hierarchy from genes to
evolution:

- Fins and limbs are homologous as paired
  appendages
- The cartilage-based endoskeleton of fish fins is
  homologous with the endoskeleton of limbs except
  for the digits, which are novel elements that arose
  during the transformation from fins to limbs
- A conserved set of cell-cell interactions between the
  epithelial covering of fin/limb buds and fin/limb
  mesenchyme initiates growth and patterning of
  fin and limb buds and of their constituent skeletal
  elements
- Conserved genes and gene networks, acting
  through these epithelial-mesenchymal interactions, initiate fin/limb differentiation, growth, and
  morphogenesis

Today, fins and limbs are studied in an amazing
variety of contexts—embryonic and postnatal development, morphology, function, repair, regeneration,
congenital diseases, evolution. Major current themes
of research within evolution are: how fins arose and
diversified among the more than 26,000 species of
fishes alive today (as well as in many groups of early
vertebrates not properly called "fishes"); how fins
were transformed into limbs at the origin of the tet-

rapods; how the limbs of some tetrapods that have readopted an aquatic existence—notably plesiosaurs and whales—modified their limbs into flippers; how the limbs of some tetrapods that took to the air— notably pterosaurs, bats, and birds—modified their limbs into wings; the mechanism(s) underlying digit or limb loss, notably in lizards, snakes, and caecilians.

Just as it did not escape Owen's attention, it has not escaped the attention of the current generation of fin and limb researchers that all of these questions can be investigated by analyzing fin and limb development. Indeed, one of the reasons for the integration of evolution and development as evolutionary developmental biology over the past two decades has been the integration of studies on the genetic and cellular regulation of fin/limb development with new fossil finds, and with such emerging concepts as constraint, modularity, conserved regulatory genes, gene cascades, levels of selection, and a hierarchical approach to homology, to name but few.[3] Were Owen alive today he would be a major contributor to all these areas—the breadth of his grasp and tenacity of his reach encompassed just as many diverse fields a century and a half ago.

The essays accompanying Owen's text position his study into several important contexts. Ron Amundson, who spearheaded this reissue of Owen's essay, takes on the topic that formed the basis of the celebrated debate between Geoffroy and Cuvier before *l'Académie royale des sciences,* in Paris, on eight Monday evenings between February 8 and April 5, 1830. That topic, put simply—whether form determines function or function form—captured the philosophical differences between

3. See Carroll (1997), Shubin et al. (1997), Carroll et al. (2005), and Hall (2005, 2007) for flavors of current research and thinking on fin/limb development/evolution.

functionalism (argument from design) and structuralism. Amundson evaluates the debates in depth and relates them to Owen's own structuralist position. As he shows, Owen's essay and his elaboration of archetypes (of vertebrates; of fins and limbs) "turned mainstream British biology to structuralism." Owen's concepts of homology, of bone-by-bone comparisons and his approach to species as variations on the archetype, brought him slowly and reluctantly toward a view of transformation in the origin of species that was anti-adaptation—pure structuralist. In transforming Owen's archetypes into ancestors, Darwin placed morphology as a central class of evidence for descent with modification. Amundson shows how much Owen's approach influenced Darwin. Owen tried to explain form, Darwin used form to explain evolution. By the end of his life, Owen wished he had done the same.

Kevin Padian, wearing his palaeontological and historical hats, discusses Owen's approach to homology and archetypes, and how Owen can be situated with respect to Geoffroy and the German, Edinburgh, and London transcendentalists (who sought explanations for form in transformation during ontogeny), Cuvier, and the Oxford and Cambridge followers of the English cleric William Paley (who stood for function over form, non-transformation, and divine creation). Padian shows just how nuanced Owen's views were—Darwin certainly did not divine them all in Owen—and how much of continental tradition was transferred to England and Scotland by Owen. This essay goes a long way toward resurrecting the influence of Richard Owen. One wonders whether, if Darwin had possessed Owen's "hubris, his pettiness, his insincerity, and his malevolence toward his rivals," Darwin's friends would have been his enemies, and we would be approaching the 150th anniversary of Alfred Russel Wallace's *On the Origin of Species.* Science is contingent on history and as deeply embedded in human frailties as are any of our other activities.

In the third essay, Polly Winsor and Jennifer Coggon analyze the frontispiece used by Owen, a winged person (angel?) about to cut the throat of (sacrifice?) a bull (cow?). These two divergent images are connected (homologized?) by numbering the homologous bones of their skeletons. In the best traditions of scientific and detective tradition, Winsor and Coggon identified the "person" as the Roman goddess Victory, who, indeed, is sacrificing a bull in celebration of a victory. They go further and use pieces of evidence from London in the 1840s to argue that Owen used the image as an advertising gimmick. Wonderful stuff.

These and so many other reasons justify the wisdom of the editor and the press in reissuing Owen's *On the Nature of Limbs,* which, with its accompanying essays, I highly recommend.

REFERENCES

Carroll, R. L. 1997. *Patterns and processes of vertebrate evolution.* Cambridge: Cambridge University Press.

Carroll, S. B., J. K. Grenier, and S. D. Weatherbee. 2005. *From DNA to diversity.* 2nd ed. Malden: Blackwell Science.

Clack, J. A. 2002. *Gaining ground: The origin and evolution of tetrapods.* Bloomington: Indiana University Press.

Hall, B. K., ed. 1994. *Homology: The hierarchical basis of comparative biology.* San Diego: Academic Press.

Hall, B. K. 2005. *Bones and cartilage: Developmental and evolutionary skeletal biology.* London: Elsevier/Academic Press.

Hinchliffe, J. R., and D. R. Johnson. 1980. *The development of the vertebrate limb.* Oxford: Oxford University Press.

Hinchliffe, J. R., J. Hurlé, and D. Summerbell, eds. 1991. *Developmental patterning of the vertebrate limb.* NATO Advanced Science Institutes Series A: Life Science. New York: Plenum Publishing Corp.

Moore, J. A. 1993. *Science as a way of knowing: The foundations of modern biology.* Cambridge: Harvard University Press.

Owen, R. 1848. *On the archetype and homologies of the vertebrate skeleton.* London: J. van Voorst.

Shubin, N., C. Tabin, and S. Carroll. 1997. Fossils, genes, and the evolution of animal limbs. *Nature* 388:639–48.

Zimmer, C., and C. Buell. 1998. *At the water's edge.* New York: Simon & Schuster.

RICHARD OWEN AND ANIMAL FORM

Ron Amundson

1. FORM VERSUS FUNCTION

Richard Owen was a central figure in a great debate within theoretical biology during the first half of the nineteenth century. The debate was not evolution versus special creation. It was *form versus function*. Are the body parts of organisms best understood in terms of the functions they serve for the organism, or does organic form reflect the operation of structural laws that are independent of functional outcomes (Russell 1916; Ospovat 1981; Padian 1995)? This debate seems strangely abstract today. But by the standards of the time it was much more concrete than the question of the origins of individual species. The evolution versus creation debate concerns what we would now call the "ultimate causes" of species origins (Mayr 1961). Scientific methodology in the early nineteenth century was very cautious about hypothetical reasoning. Speculation about unobserved causes was discouraged. Unlike the question of species origins, the form-function debate concerned observable body parts, not historically remote events. The discovery of law-like generalizations about observable bodies was thought to be a more suitable goal for a conscientious scientist than the quest for ultimate causes. *Functionalism* (or teleology) versus *structuralism* (based on comparative morphology and embryology) was the hotly contested issue. Functionalism dominated during the early part of the century under the influence of Baron George Cuvier. Functionalism was congenial to special creationism, because the Argument from Design (the centerpiece of natural theology) was associated with a functionalist reading of biology. This approach was the theme of the eight volumes of the *Bridgewater Treatises,* published in the 1830s. Cuvier died in 1832. The openly religious goals of the *Bridgewater Treatises* may have harmed the scientific status of functionalism. In any case, functionalism was beginning to seem old-fashioned by the 1840s.

Structuralism had originated on the European continent, and in Britain was initially associated with social radicalism (Desmond 1989). Richard Owen was a member of the conservative mainstream of the British scientific establishment. Owen's writings of the 1840s, culminating in *On the Nature of Limbs,* turned mainstream British biology to structuralism. This achievement involved several steps. First, Owen convincingly argued that functionalism (understood as teleology at the time) failed to account for the facts of biology. Second, he reformulated structuralism to better accord with British empiricist methodology, and argued that it was consistent with a pious attitude towards religion. The third step was controversial at the time, and has often been overlooked in the modern day. Owen argued that structuralist biology, properly understood, could contribute towards an understanding of the natural causes of species origins.

Owen did not offer a theory of species origins.[1] But he did imply that the achievements of structuralist biology (including the Vertebrate Archetype elaborated in *Limbs*) could point the way toward such a theory. And he was correct. Charles Darwin misread Owen's views regarding species origins. In the *Origin of Species,* Darwin mistakenly claimed that Owen was committed to the immutability of species (Darwin 1859, 310). Nevertheless, Darwin used Owen's results to great effect in arguing for the fact of evolution.

Many of the historical writings of the late twentieth century describe Owen as a reactionary influence on evolutionary biology. This view is primarily based on Owen's reputed Platonism and idealism, which are claimed to be inconsistent with evolutionary understanding. His version of structuralist biology has been widely reported as an "idealistic version of the Argument from Design." This interpretation, which originated with Peter Bowler (1977), was followed by such critics of structuralism as Ruse (1979) and Mayr (1982), but also by such sympathizers with structuralism as Ospovat (1981) and Gould (2002). It is a misunderstanding, caused (I believe) by twentieth-century commitments to neo-Darwinian evolutionary theory. Owen's view of evolution is indeed very different from Darwin's. But current developments in evolutionary biology, especially in evolutionary developmental biology (evo-devo), make Owen's work relevant today in a way that it was not in the 1970s and 1980s. This essay will examine Owen's structuralism, his views on spe-

1. Finally in 1868 Owen did propose an evolutionary theory he called "the derivative hypothesis" (Rupke 1994, 248). His earlier speculations are thought to have centered either on the phenomenon of alternation of generations, or on teratology (Richards 1987, 1994).

cies origins, the relevance of idealism and Platonism, and Darwin's own use of Owen's work. It will conclude by considering the differences between Owen's and Darwin's approaches to the understanding of evolutionary change.

2. HISTORY OF THE DEBATE

Functionalism held that the characteristics of organisms were those that were needed in order to function. Cuvier was the early champion of functionalism, and was revered by the British natural theologians. He claimed that all similarities among organisms stemmed from commonalities of function. Requirements of function were termed the *Conditions of Existence.* Cuvier was the greatest of comparative morphologists, and established "affinities" among body parts of taxonomically related species. But, like the Bridgewater authors, Cuvier was unwilling to accept affinities that were based on structural similarities alone (Appel 1987, 45). For Cuvier and the Bridgewater authors, all *real* similarities were traceable to common function. Taxonomic classification was simultaneously functional classification.

Structuralism arose on the continent among a number of highly speculative thinkers—at least that was the assessment of most British empiricists. Structuralist thinkers included the *Naturphilosophen,* who were considered overspeculative even by their continental successors (Nyhart 1995). They also included Johann Wolfgang von Goethe, Karl Friedrich von Baer, and Étienne Geoffroy St. Hilaire. A public confrontation between Cuvier and Geoffroy in 1830 was the first dramatic clash between functionalism and structuralism (Appel 1987). Even though continental structuralists held a wide variety of metaphysical and theoretical commitments, they were generally characterized as idealist and transcendental; as a group they were known as *transcendental anatomists* or *idealistic morphologists*.[2] These labels are explained in part by the strong influence of Kantian methodology on morphological studies (Lenoir 1982; Sloan 1992; R. Richards 2002). But the labels are also due to a generally accepted epistemological principle of the time that I term the *empirical accessibility of function* (Amundson 2005).

Before Owen's work of the late 1840s it was generally held that biological functions were accessible to observation, but that purely structural (nonfunctional) correspondences were not. The structuralists claimed that *structural* correspondences obtained between species (and also within the repeated elements of an organism's body, as we will soon discuss). Because these correspondences could not be reduced to function, their recognition was considered to be "ideal" and (by empiricists) speculative. One example of the conflict involved the furcula, a bone that Cuvier had identified in birds as a functionally important component in flight. The structuralist Geoffroy claimed to have found a corresponding bone in fish. Because the fish bone could not possibly function for flight, Cuvier refused to associate the fish "furcula" with the bird furcula. Historian Toby Appel reports Geoffroy's position to be that "the

2. Surprisingly, one of the Bridgewater Treatises did discuss transcendental anatomy approvingly. Peter Mark Roget's Treatise reported a sanitized version of transcendental anatomy in an apparent attempt to defuse its radical implications (Desmond 1989, 229; Roget 1834). Nevertheless, Roget's primary religious conclusions regarding species fixism and God's design were based not on transcendental anatomy but on Cuvierian functionalism (Amundson 2005).

furcula was not a bone specifically designed by the Creator to aid birds in flight, but rather an abstract element of organization which could serve multiple functions as it was placed in different circumstances" (Appel 1987, 87). Why should a bone be considered an "abstract element"? Because Geoffroy claimed an affinity between bones in two species, even though the two bones did not serve corresponding functions. According to the standards of the time, this amounted to an *idealization.* In order to identify common types successfully (the argument went), those types must be hypothesized *a priori,* and such an *a priori* assumption of common types constituted idealism.

Functionalists regarded such hypotheses to be contrary to empiricist principles, and stuck to the functional characterizations that were thought to be directly observable. Geoffroy proposed *Unity of Type* as a structuralist slogan to contrast with Cuvier's functionalist Conditions of Existence. Conditions of Existence versus Unity of Type was function versus structure, one of the liveliest controversies of the era. It formed the background of Owen's work, and (in a different way) of Darwin's as well.

In the two decades between the Cuvier-Geoffroy confrontation and the publication of Owen's *Limbs,* functionalism declined and structuralism gained in scientific prestige, although Britain lagged behind the continent. Continental morphology gained adherents slowly in Britain, at first among radicals. Owen's *Limbs* was one of the key factors that brought mainstream British biology in line with the continent. It did so in part by expressing structuralist biology in a conservative way.

Owen's conservatism had two aspects. The first was dutifully religious: he padded his structuralist conclusions in pious rhetoric (though this rhetoric did not

dilute his radical conclusions). The second was epistemological: Owen did his best to present his views as
having been arrived at on the basis of good British
empiricism and inductivism. He acknowledged the continental morphologists for their ideas, but blamed them
for their speculative excesses and pointed out empirical errors. He admitted that very little morphology
had been done in Britain (*Limbs*, 4), but was careful
to list the speculative flaws in the work of continental
morphologists (*Limbs*, 41, 81). This had to be done in
order to overcome the principle of the empirical accessibility of function. Owen insists that the Vertebrate
Archetype and homologies are "no mere transcendentalist dream, but true knowledge and legitimate fruit
of inductive research" (*Limbs*, 70). Radicals and lesser
figures had already argued the point, but it was Owen
who brought the argument home, and made structuralism palatable to mainstream British science.

3. RICHARD OWEN'S STRUCTURALISM

Owen had ties to two distinct groups of patrons, and
his career depended on the patronage. One was the
traditional Oxbridge establishment, consisting largely
of conservative Bridgewater functionalists and their
colleagues. Many of these were well acquainted with
the structuralist developments on the continent, and
disapproved strongly. Among the best-informed was
the surgeon Charles Bell, author of the Bridgewater
Treatise *The Hand: Its Mechanism and Vital Endowments* (Bell 1833), a source that Owen is careful to cite
whenever possible in *Limbs*. Bell was a deeply committed functionalist. His Treatise is filled with criticisms
of the "lovers of system" whose structuralist theories
are "a means of engaging us in very trifling pursuits—

and of diverting the mind from the truth." The truth is that adaptation dominates the organic world (Bell 1833, 40; for discussion of Bell see Amundson 1996).

Owen was also influenced by a London circle of intellectuals who surrounded the transcendental poet Samuel Coleridge. This group was less antagonistic to continental thought, with its idealist and pantheist overtones. Owen claimed to have reconciled the two sides of the form versus function debate (*Limbs,* 34, 84; for conciliatory statements). It is true that he practiced both kinds of work, and functionalism dominated his early research (Rupke 1994, 117). Nevertheless, his conciliatory rhetoric and his conservative patrons notwithstanding, Owen's major achievements and his deepest commitments were in structuralist theory. During the 1840s he organized and finalized the results of transcendental anatomy in a far more conclusive way than had its continental practitioners. This occurred in three steps.

(A) Analogy and homology

First, Owen clearly distinguished between *analogy* and *homology:*

> Analogue.—A part or organ in one animal which has the same function as another part or organ in a different animal. (Owen 1843, Glossary, 374)

> Homologue.—The same organ in different animals under every variety of form and function. (Ibid., 379)

A distinction similar to this one had been implicit in earlier writings of continental morphologists and others. One term (sometimes affinity, sometimes homology) was taken to designate the deep and meaningful similarities between organisms—those that revealed the underlying Unity of Type. A contrasting term indicated superficial similarities. Darwin was unfamiliar with Owen's 1843 distinction when he composed his unpublished *Essay* of 1844, and referred to the "ill-defined distinction between true and adaptive affinities" (Darwin 1909, 215). Darwin knew that Unity of Type offered something to his evolutionary views, but in 1844 he was not quite sure what that was. Owen's definition standardized the terminology, and stipulated that many superficial resemblances reflected functional similarities. Owen's structuralism is revealed even at this early stage of the discussion. Homology is based on structure (not function), and it provides the deepest insights into organic nature. Darwin had been skeptical about "true affinities" in his *Essay*. But after recognizing Owen's clarification (true affinities were precisely homologies), Darwin discussed homology extensively throughout the *Origin*.

(B) Homologies of all elements of the skeleton

Owen's second step was to document the fact that the *entire skeletons* of vertebrate groups could be shown to correspond, bone for bone, with those of other vertebrate groups. Previous morphologists had mostly been content to search for surprising correspondences, such as the homologues of mammalian ear ossicles in non-mammal groups. (During the 1830s Geoffroy had identified them with the opercular bones of fishes, Riechert with the jawbones of reptiles; Riechert's identification is still regarded as correct.) In *The Archetype and Homologies of the Vertebrate Skeleton* (1848), Owen catalogued the names by which various vertebrate bones had been designated by the specialist-anatomists who had named them. Anatomists who specialized in different animals (for example birds, fish, horses, and humans) used different names for bones, often reflecting a superficial resemblance that the bone possessed in their group. Owen tabulated the various names and descriptions by which all of the bones were (separately) known. He then assigned the corresponding bones the same name, in many cases replacing a long anatomical description with a brief name. Plate I in *Limbs* contains a numbered list of these names, and the numbers are used throughout the book to designate the homologous bones in all species under discussion. Owen described his renaming project as if it were a simple matter of commonsense pragmatism and British empiricism. He stated a Lockean philosophy of naming: names should be arbitrary symbols; they should not be imbued with theoretical baggage that might retard future research in science. He describes the project as if it is a simple empirical catalog with no theoretical ambitions or presumptions. But this is a strategic posture, aimed at circumventing the principle of the empirical accessibility of function. Owen disguises one huge theoretical assertion as a simple empirical fact. The assertion appears in this passage: "To substitute names [of bones] for phrases is not only allowable, but I believe it to be indispensable to the right progress of anatomy; but such names must be arbitrary, or at least, *should have no other significantion than the homological one*" (Owen 1848, 3; emphasis added). The underlying theoretical assumption of the entire project is that all vertebrates are built on a single body plan. Homologies are the elements of this body plan. This is a direct contradiction of the classificatory doctrines of Cuvier and the Bridgewater natural theologians. Recall that Geoffroy's identification of the furcula bone in birds and fishes had earlier been labeled an idealization! Owen uses the exact same method, but does so in such a brisk, no-nonsense manner that the radicalness of the project was unnoticed. The practice that Bell had condemned as "trivial pursuits," and that others suspected of various sins, from idealism to pantheism to atheism, was being repackaged by Owen into an apparently harmless form.

(C) Types of homology

Owen's third step was to articulate three distinct kinds of homology: serial homology, special homology, and general homology. *Serial homology* was the relation among repeated elements in an individual body. Examples are the relation between forelimbs and hindlimbs, and among successive vertebra (*Limbs,* 21). *Special homology* (the kind discussed in preceding paragraphs) indicates a correspondence between body parts of different species. *Limbs* gives dramatic illustrations of the homological nature of the differently adapted limbs of vertebrate species (*Limbs,* 5–8).

General homology is a subtler and more difficult relation, and is seldom accurately reported in modern discussion (except Camardi 2001). General homology is based on Owen's view that vertebrates are segmental organisms. A "vertebra" for Owen is not simply a bone, but an entire bodily segment. Each of these segments is itself made up of parts that stand in definite relations to one another within the segment. Owen illustrates the relation between the ideal vertebra (as a segment made up of elements) and a natural "vertebral segment" in Figures 7 and 8 (*Limbs,* 42, 43). The natural segment in Figure 7 is a bird thorax, including not only the bones commonly named "vertebrae" but also the ribs and sternum. The schematic representation of the generalized ("ideal") vertebral segment in Figure 8 includes the neural arch above and the haemal arch below, and each of the elements is named (neurapophysis, haemapophysis, etc.). Each vertebrate's body is made up of a series of these segments, and each segment contains the same elements. The goal of the study of general homology is to identify the bones of natural organisms both with respect to (a) which vertebral segment the bone belongs to in the series of segments that composes the body, and (b) which *element of the segment* (neurapophysis, haemapophysis, etc.) the bone represents in its respective segment. So we see that *Limbs* depicts not one archetype, but two. They are the Vertebrate Archetype (Plate I) and the "Ideal vertebral segment" (Figure 8). The former is a serial construction of the latter.

The importance of Owen's concept of general homology is best illustrated in his discussion of the vertebrate skull. Like almost all of his contemporaries, Owen accepted the *vertebral theory of the skull.* This view interprets the skull as a series of modified vertebrae (vertebral segments for Owen). The most direct description of the view occurs on page 43 of *Limbs,* where Owen explains how the ideal vertebral segment is modified differently in the skull and in the thorax. Cranial vertebrae have enlarged neural arches, to contain the brain, whereas thoracic vertebrae have enlarged haemal arches to contain the organs of circulation. The view of vertebrates as segmental animals implies that the skull itself is made up of modified vertebrae.[3] Owen's first discussion of general homology in *Limbs* makes use of the vertebral theory of the skull. He reports that "the basilar part or process of the occipital bone in human anatomy is the 'centrum' or body of a cranial vertebra" (*Limbs,* 4). The occipital bone is a part of the last vertebral segment of the

---

3. Although the vertebral theory of the skull has been abandoned, its significance should be recognized. A morphologist cannot simply assume that vertebrates *somehow* acquired heads: the existence of the head must be given a morphological explanation. Ideally, the head should be understood in a way that reflects its unity of structure with the rest of the body.

skull, the so-called occipital vertebra. Owen goes on to describe how this implies that the item is therefore not really a "process" (an extension of a separate bone), but an independent vertebral element, *serially* corresponding with the centra of all other vertebra in that animal's body, and *specially* corresponding to independent bones (not to processes) in the bodies of cold-blooded vertebrates. It also has a special developmental relation to the notocord (*chorda dorsalis*) that identifies it as the centrum of that vertebral segment. So the general homology of this bone extends in three dimensions: serially in the animal's body, specially in its relation to homologues in other vertebrates, and developmentally in the centrum's special relation with the notocord during embryogenesis. The complexity of this illustration shows the importance of general homology to Owen. It would be central to his thought on species origins.

4. OWEN ON SPECIES ORIGINS

After Lamarck, the problem of species origins was in the background during the early part of the century. It burst into public attention with the anonymous publication of *Vestiges of the Natural History of Creation* (Chambers 1844). The book was popularly written. It included a wide range of scientific reports that the author claimed to supported evolutionary conclusions, but it was clearly not written by a professional naturalist. Most scientists condemned it, and it was a raging popular success. The author of the *Vestiges* was not the only person to consider the possibility of a naturalistic cause of species origins—he was just the only one to claim that he knew the cause (the "law of development," said to be just as powerful as the "law of gravitation"). Many others had a general belief in naturalistic causes of species origins but had no idea what those causes were, and so declined to speculate about them.

In contrast to the conservative and special creationist Bridgewater authors, liberals on species origins included Charles Lyell, Baden Powell, and William Carpenter (Ruse 1979; Corsi 1988; Rupke 1994). The successive appearance of species in the geological record was an observed fact. Nevertheless, species fixism was also considered a well-supported scientific fact.[4] This left the liberals in a quandary. Their naturalism about species origins was based on a general commitment to exceptionless natural laws, and an opposition to miracles. The liberals were theists (or perhaps deists). They believed that God had originally created the world with its full set of exceptionless natural laws. These laws were called *secondary causes* to discriminate them from God's original creation, the First Cause. This was a time of cautious empiricism in scientific method, and it was considered perfectly appropriate to remain silent regarding matters on which one lacked sufficient evidence to support a hypothesis (Hull 1973, 1983; Ruse 1979). Those who accepted naturalistic causes of species origins were not inclined to speculate about what those causes were. This reticence may have been influenced by a fear of confrontation with religious conservatives, but it was certainly influenced by the empiricist caution of the period.

Richard Owen cautiously began to discuss natural-

4. The common opinion is that pre-Darwinian species fixism was based on metaphysical essentialism. This is inaccurate. Empirical grounds had been gathered for species fixism ever since the work of Linnaeus in the middle of the eighteenth century (Müller-Wille 1995; Amundson 2005). Pre-Darwinian systematists were not essentialists (Winsor 2003).

istic species origins in *Archetype,* and expanded his comments in *Limbs.* His remarks were guarded, and surrounded by pious rhetoric. This may have been why Darwin and some others didn't even recognize them. But Owen's conservative allies recognized them, and attacked them harshly.

Owen's comments on species origins are closely intertwined with his concept of general homology and its role in the embryological formation of bodies. Development occurs under the influence of two general principles (or laws or forces). One force is responsible for Unity of Type, and the other for diversity and adaptation. The structural force tends to produce the repetition of identical elements in a body. It was originally described as a "polarizing force" such as that involved in magnetism or the growth of a crystal (Owen 1848, 171). In the vertebrate body it produces repeated vertebral segments along the front-to-back axis, the prime example of serial homology. The structural force also produces special homology, the identity of body parts between species (e.g., vertebrate limbs). The repeated elements produced by the structural force are differently modified by the adaptive force to serve diverse functions. These two forces act, to some extent, in opposition. The structural force dominates in lower life forms such as worms and starfishes. It also dominates in the lower forms within a class; the lowest vertebrates are closest to the Archetype (*Limbs,* 49, 59).

Owen's two forces, structural and adaptive, account for both the diversity among species and the variation in parts of the body of a single organism.[5] Vertebrate bodies have diverse, specialized segments even though

---

5. "Account for" in an abstract sense, of course, suitable to Owen's structuralism though not for Darwinian styles of evolutionary explanation.

the segments are ideally identical. Both the adaptive and structural forces are at work *during the embryological development of the individual organism.*[6] These two forces are also responsible for the unity and diversity that we see between species. The homological identity of vertebrate limbs results from the similar action of the structural force acting during the embryological development of the dugong, the mole, the bat, the horse, and the human. These limbs are adaptively distinct because the adaptive force works throughout embryological development.

Owen's concept of general homology led him to recognize a similarity between the adaptive specialization of segments in an individual body (a skull segment and a torso segment) and the adaptive specialization between different species of a common type (a bat's forelimb and a horse's forelimb). Just as an individual vertebrate body is a series of variations on the theme of the ideal vertebra, distinct species are variations on the Vertebrate Archetype. In Owen's view, the recognition of these patterns, and the structural and adaptive forces that produced them, could lead to an understanding of the origins of new species on earth. This is not an "evolutionary theory" in the Darwinian or Lamarckian meaning of the term. But it can be seen as a step toward such a theory.

Here is Owen's expression of that point in *Archetype:* "To trace the mode and kind and extent of modification of the same elementary parts of the typical segment throughout a large natural series of highly organized animals like the vertebrata; . . . is one of the legitimate courses of inquiry by which we may be

6. The two principles operate "in the development of an animal body . . . during the building up of such bodies . . . in the arrangement of the parts of the developing frame" (Owen 1848, 171–72).

permitted to gain an insight into the law which has governed the successive introduction of specific forms of living beings into this planet" (Owen 1848, 106). In *Limbs,* Owen's discussion of naturalistic causes of species origins occurs prominently at the very end. The passage begins, "To what natural laws or secondary causes the orderly succession and progression of such organic phaenomena may have been committed we as yet are ignorant. But if, without derogation of the Divine power, we may conceive the existence of such ministers . . . " (*Limbs,* 86). Owen is clearly not a conservative creationist regarding species origins. He is pious, but he assumes the existence of secondary causes (natural laws) that brought new organic forms into being. He asserts that the study of homology will give us new insights into those laws.

The conservative backlash against Owen's work of the 1840s was focused both on the passages that discuss natural laws of species origin, and on his characterization of the two forces said to be responsible for the forms of bodies. His description of the forces changed in the three years between the original publications of *Archetype* and *Limbs.* Of special interest is Owen's reference to Platonic idealism. Neo-Darwinian commentaries often emphasize Owen's Platonism as an indication of the spiritualistic basis of his thought. But the details of Owen's use of Platonism reduce its metaphysical significance.

In *Archetype,* the structural force is described in virtually materialist terms. It is an "all-pervading polarizing force" (Owen 1848, 171). Owen's use of the "polarity" concept aligns the vertebrate body, and its front-to-back bodily axis of segmental repetition, with contemporary explanations of phenomena like crystallization and magnetism. The adaptive force corresponds to "the *ideas* of Plato . . . which [Plato] defined as a sort of models, or moulds in which matter is cast, and which regularly produce the same number and diversity of species" (Owen 1848, 172). Structure is materialist in origin, and adaptation is Platonic. But in *Limbs* the story suddenly changed. Owen identifies the *structural* force (not the adaptive force) as Platonic. The nature (or signification, or *Bedeutung*) of a limb is "that essential character of a part which belongs to it in its relation to a predetermined pattern, answering to the 'idea' of the Archetypal World in the Platonic cosmogony" (*Limbs,* 2–3). In *Archetype* the adaptive force was Platonic, but in *Limbs* the structural force is Platonic. Owen gives no explanation for this flip-flop. But recent historical sleuthing has uncovered a probable cause. Owen reversed his Platonism in response to a challenge from the Oxbridge conservatives, who disapproved of Owen's willingness to explain Unity of Type by natural causes.

Owen's Platonic reversal followed a letter he received from the Cambridge conservative William Conybeare in 1848. Conybeare suggested that Platonic idealism be relocated in Owen's theory, to replace materialist polarity as the structural force:

> [Plato] meant the archetype forms of things, as they existed in the creative mind. . . . To me the true . . . analogy seems to be the mind of a manufacturer about to produce his work; a shipwright his ship—an instrument maker his piano, or organ. (Coneybeare quoted in Rupke 1994, 202)

With the background of Conybeare's letter, Nikolaas Rupke interpreted Owen's 1849 redeployment of Platonism as an attempt to "placate the powerful Oxbridge faction among Owen's supporters" (Rupke 1994, 204). Unity of Type had for years been a thorn in the side of adaptationist natural theology because of its apparent *lack of need* for theological under-pinnings. Owen's polarizing force simply drove the thorn deeper. But if Owen could be convinced to relo-cate his Platonism to Unity of Type, transcendental anatomy might be brought under Plato's supernatural supervision. Owen cooperated.

This concession was not Owen's only response to Conybeare's letter, however. *Limbs* did Platonize the Archetype, and Owen made use of the idealist rhetoric to express his piety. But Owen's real target of criticism in both books was teleological biology as it had been practiced by Cuvier and the Bridgewater authors. He was championing structuralism; let Platonism fall where it may. Conybeare had likened the Archetype to a plan in the mind of a designer, "a manufacturer about to produce his work; a shipwright his ship." Owen's let-ter in reply to Conybeare explains that Conybeare's analogy was mistaken. The structural facts of anat-omy could not be accounted for by analogy to "works of human art" (Sloan 2003, 60–61). Unity of type is *not* like an intelligent designer's plan, and merely attach-ing Plato's name to it doesn't change that fact. Owen did use Conybeare's shipwright analogy in *Limbs,* but he stood it on its head. This almost satirical twist on the shipwright analogy will be discussed in the follow-ing section.

The statement about species origins at the end of *Limbs* is prefaced by a religiously motivated discus-sion that is difficult for modern readers to interpret. We are far too eager to read special creationism and other religious question-begging into passages that were only intended to express piety, not a scriptural grounding for scientific results. But I submit that the pious tone in these passages can be easily separated from Owen's actual assertions. In the most directly religious passage, Owen claims that his "Platonic" views on the Vertebrate Archetype refute an obscure argument for atheism attributed to ancient Greek atomists. Shades of Bridgewater! But let us read carefully what he claims to have proven: "[T]he knowledge of such a being as Man must have existed before Man appeared. For the Divine mind which planned the Archetype also foreknew all its modifications" (*Limbs*, 85–86). This certainly affirms Owen's theism. But it does not implicate special creation. Liberal theists all believed in God's original creation of the world, along with its secondary causes. They also believed that God knew ("foreknew") the eventual effects of these secondary causes. God created the force of gravity, and so foreknew the paths that the planets would take. For Owen, the Vertebrate Archetype was somehow involved in the secondary causes for species origins. God established those secondary causes, and of course foreknew their consequences. This does not imply that Man (or any other species) was created individually by God. Even in the midst of his rotund piety, Owen leaves ample room for natural species origins.

Owen's discussions of the natural origins of species were guarded. But they occurred in a crucial location. They were embedded in discussions of the Unity of Type, and the forces that controlled organic form both within and between vertebrate bodies. The mention of naturalistic species origins *in the context of* Unity of Type was a bold step. It was too bold for many conservative critics. Adam Sedgwick criticized Owen in print the following year, expressing grave doubts about Owen's pantheist tone. The *Manchester Guardian* angrily condemned Owen in an editorial for his theologically unacceptable expression of "what is called the theory of development" (E. Richards 1987, 163ff.). Unlike Conybeare's letter, these were public condemnations. Stung, Owen ceased to publish on species origins. By the time he returned to the subject, he had already been scooped by Darwin's *Origin*.

5. ANTI-ADAPTATIONISM

On the Nature of Limbs is most frequently cited for two things: the Vertebrate Archetype and Owen's alleged Platonism. I consider the Archetype very important, and Owen's Platonism a complete red herring.[7] A third feature of the book should not be overlooked. *Limbs* is intensely structuralist. It gives example after example of the failures of adaptationist, teleological reasoning. If one were to search for comparisons to *Limbs,* the Bridgewater Treatises are the last place to look. A far closer similarity is "The Spandrels of San Marco and the Panglossian Paradigm" (Gould et al. 1979). Owen was certainly not as iconoclastic as Gould and Lewontin, but his defense of structuralism and critique of adaptationism was every bit as intense. The first refutation of adaptationism in *Limbs* includes Owen's reversal of Conybeare's adaptationist shipwright analogy.

Owen first illustrates homology with detailed illustrations of the structures of vertebrate limbs: the mole, dugong, horse, bat, and human (*Limbs,* 4–9). Limbs are differently adapted, but share a common

7. Rupke points out that Owen's Archetype is unlike a Platonic Form (Rupke 1993, 243). Platonic Forms were taken to be the highest and most perfect exemplar, whereas Owen's Archetype is the most general and undifferentiated form. Given his flip-flop on its application, I see so little evidence of the metaphysical significance of Owen's Platonism that its deep interpretation is inconsequential. His idealism is no more metaphysical than Geoffroy's: a willingness to hypothesize beyond the limits of the principle of the empirical accessibility of function.

structure. His point is that the commonality of structure is *not* traceable to function—a direct denial of the natural theologian's argument. Owen proves his point by using a piece of standard natural theological rhetoric, then standing it on its head. He compares limbs with human transportation inventions. His very first example is Conybeare's shipwright: "To break his ocean bounds, the islander fabricates his craft, and glides over the water by means of the oar, the sail, or the paddle-wheel" (*Limbs,* 9). Owen proceeds to list a number of other human transportation devices. He then departs from the natural theological game plan.

Instead of pointing out similarities between natural bodies and human inventions, he indicates a crucial *difference* in the cases. Human ingenuity adapts an invention *directly* to its purpose, and does not make modifications to a common plan in order to produce a new invention: "There is no community of plan or structure between a boat and a balloon" (*Limbs,* 10). Unity of type in the organic world does not correspond to a functional plan in the mind of a human inventor. The deepest truths in the organic world are those of Unity of Type and homology; these truths cannot be explained by teleological reasoning. Conybeare's adaptationist shipwright is held up as the very model of adaptationist foolishness.

Other challenges to adaptationism take a slightly different form. Owen considers an adaptationist explanation of a morphological feature in humans. He then proves that other species share that same morphological feature, but do not share the functional need implied by the adaptationist explanation. The first example is the first digit on the human foot. The hallux, or big toe, has fewer phalanges than the other four digits even though it is larger. The adaptationist would explain the reduced number of phalanges by the need for strength in the big toe in its use in human walking. Sounds fine, as long as we look no further. But Owen immediately points out the failure of this explanation: the elephant and the seal also have reduced numbers of phalanges in their first digit, even though that digit is identical both in size and function to the other digits on the animal's limb (*Limbs,* 37).

A second example of this critique involves the complicated pattern of unfused bones in the skull of an infant human. It had been proposed that the pattern of incompletely formed bones was an adaptation for the passage of the large human skull through the birth canal. Indeed the infant skull's compressibility does aid in birth. Nevertheless, "the same ossific centers are established, and in similar order, in the skull of the embryo kangaroo, which is born when an inch in length, and in that of the callow bird that breaks the brittle egg" (*Limbs,* 40). The "principle of special adaptation" (i.e., adaptation in the individual species) fails to explain these structures. The centers of ossification are homologous. They are not confined to one species, nor are they confined to the species that have a functional need for them. Adaptationism fails again.

The careful reader will notice that each of these critiques is followed by a pious passage in which Owen assures his reader that his skepticism about teleology does not imply irreligion. This was partly a sign of the times, but partly idiosyncratic to Owen. References to religion were common in popular scientific writings, even when the topic was not concerned with potentially controversial topics. But Owen is especially eager to demonstrate his piety in *Limbs.* He was aware that the Bridgewater conservatives had aligned structuralism with irreligion, and that his own writings were being monitored by them. Nevertheless, if the pious passages are read carefully, they are consistent with naturalistic species origins. Consider the pious conclusion to the skull discussion: "[I]f the principle of special adaptation fails to explain [the homologies], and we reject the idea that these correspondences are manifestations of some archetypal exemplar on which it has pleased the Creator to frame certain of his living creatures, there remains only the alternative that the organic atoms have concurred fortuitously to produce such harmony" (*Limbs,* 40). The modern reader will see this as a false dichotomy. Surely special adaptation, divine archetypes, and sheer accident are not the only possible explanations. Ordinary natural laws are another! But I propose that ordinary natural laws (secondary causes) are precisely how Owen thought that the "archetypal exemplar" became "manifested" in the world (see again the concluding passage from *Limbs* on page 86). The divine creation of the archetype at the beginning of time corresponded to the divine creation of the law of gravity. God, the First Cause, created both the law of gravity and the organic laws by which individual species came into existence. Owen was pious in *Limbs,* even to the point of obfuscation. But he was not a special creationist.

6. DARWIN'S USE OF OWEN

If one fact about Charles Darwin is beyond question, it is that he was an adaptationist. This can be seen in Darwin's earliest speculations on species change. In his 1837 Notebook B he wrote, "The condition of every animal is partly due to direct adaptation & partly to hereditary taint" (Barrett et al. 1987, 182). Even at this early stage, Darwin saw the true nature of the species as embodied in its environmental fit; its hereditary structure was mere "taint." He remained an adaptationist even as the scientists around him in the 1840s were turning away from adaptation and towards structure. Owen's *Limbs* was a strong influence on this movement in Britain. Owen claims that homologues are "definable and recognizable under all their teleological modifications, . . . through every adaptive mask" (*Limbs,* 41). For the structuralist Owen, adaptation was a "mask." For the adaptationist Darwin, hereditary structure was a "taint." Structure meets function.

Nevertheless, Darwin studied the structuralists and made good use of their work. Many commentators consider Darwin's theory to have been essentially complete in his *Essay* of 1844. Ospovat, on the other hand, argues that the morphological and embryological research that Darwin read after 1844 was very important to the strength of his book, especially to his argument for the fact of common descent. This included the work of Owen, and especially *Limbs.* Darwin's own copy of *Limbs* has this marginal note: "I look at Owen's Archetypes as more than ideal, as a real representation as far as the most consummate skill and loftiest generalization can represent the parent form of the Vertebrata" (quoted in Ospovat 1981, 146). This reinterpretation of Owen's archetype as an ancestor was a very important step in Darwin's use of morphology to argue for common descent.

The section on morphology in chapter 13 of the *Origin* acknowledges the work of Owen and other transcendental anatomists. Darwin begins with homologies of the vertebrate limb, and points out that the same names can be given to bones in widely different animals. (Darwin almost certainly adopted this point from Owen's *Archetype,* its most dramatic statement.) His following two paragraphs allude to points from Owen's *Limbs,* although the spin is Darwin's own.

First, on teleology: "Nothing can be more hopeless than to attempt to explain this similarity of pattern in members of the same class, by utility or by the doctrine of final causes. The hopelessness of the attempt has been expressly admitted by Owen in his most interesting work on the 'Nature of Limbs.' On the ordinary view of the independent creation of each being, we can only say that so it is;—that it has so pleased the Creator to construct each animal and plant" (Darwin 1859, 435). Darwin clearly credits Owen with proof of the failure of teleology in this passage, but he does so in a slightly backhanded way. I invite the reader to notice that Owen didn't just "admit" the failure of teleology in *Limbs:* he gleefully proved it! Darwin gains a definite advantage from Owen's point, as can be seen by comparing the 1859 passage in the *Origin* to the essay of 1844. In the earlier work, Darwin had recognized that Unity of Type *can* be explained by common descent. But he had not recognized (at least not openly) that Unity of Type *cannot* be explained adaptively! As an adaptationist himself, he may not have been on the lookout for the breakdowns of adaptationism. Owen, however, was on the lookout. His success shows Darwin how to refute one particular brand of adaptationism, the brand that special creationists had relied upon. Owen (and Darwin following him) refutes *special* adaptationism, the kind that is applied directly to species without recognition of the constraints of common descent.[8]

8. It might be thought that Darwin was a "special adaptationist" also, in that natural selection works within a species (or population) and not at higher taxonomic levels. While this is true, traits produced by natural selection are not restricted to the species level for Darwin. They are generally passed on to daughter species, and can even become a part of the "type" (Darwin 1859, 206). So the

The next paragraph contains Darwin's flourish of transforming the archetype to an ancestor: "If we suppose that the ancient progenitor, the archetype as it may be called, of all mammals, had its limbs constructed on the existing general pattern, for whatever purpose they served, we can at once perceive the plain signification of the homologous construction of the limbs throughout the whole class" (ibid.). Owen was so strongly associated with the concept of the archetype by this time that there was no need to cite him. Darwin did, of course, consider Owen committed to species fixism when he wrote the first edition of the *Origin*. So his use of the expression "plain signification" may have a jibe. Signification had been the term Owen used at the beginning of *Limbs* to translate the German *Bedeutung,* a term that he had borrowed from Lorenz Oken's first publication of the vertebral theory of the skull (Oken 1807). Owen thought that the *Bedeutung* of limbs was their position in terms of general homology; Darwin believed it simply to be the ancestor's limb.

Darwin next discusses serial homologies, including the vertebral theory of the skull. On this topic he again deploys one of Owen's critiques of teleology, directly citing Owen's refutation of the teleological explanation of infant skull structure: "Why should the brain be enclosed in a box composed of such numerous and such extraordinarily shaped pieces of bone? As Owen has remarked, the benefit derived from the yielding of the separate pieces in the act of parturition of mam-

---

existence of an adaptive trait does not demonstrate that the trait was created in and for the species that holds it. This is why special adaptationism is an argument for special creationism but Darwin's adaptationism is not.

mals, will by no means explain the same construction in the skulls of birds" (Darwin 1859, 437). Then Darwin presents his version of Owen's interaction of the structural (vegetative) force with the adaptive force. Darwin subsumed the structural force under heredity and the adaptive force under natural selection: "An indefinite repetition of the same part or organ is the common characteristic (as Owen has observed) of all low or little-modified forms; therefore we may readily believe that the unknown progenitor of the vertebrata possessed many vertebrae; . . . consequently it is quite probable that natural selection, during a long-continued course of modification, should have seized on a certain number of the primordially similar elements, many times repeated, and have adapted them to the most diverse purposes" (Darwin 1859, 437–38). Owen's principle of vegetative repetition followed by adaptive specialization has always been an important part of morphology. But it has taken on a new importance in evo-devo, with crucial aspects of evolution attributed to duplication and subsequent specialization of entire gene families (Gerhardt et al. 1997). Several neo-Darwinian commentators have discussed Darwin's comments on Owen in the morphology section of the *Origin,* and have shown great reluctance to accept that Darwin was significantly influenced by Owen. Ernst Mayr interprets Darwin's rejection of teleology (quoted above) as a criticism of the transcendental anatomists themselves, precisely the reverse of my reading (Mayr 1982, 464). David Hull discusses a passage that Darwin later added in the fourth edition to the end of the paragraph: "But this is not a scientific explanation." Hull reads this as a rejection of Owen; I read it as a rejection not of Owen but of special creationism (which Darwin by then knew was not Owen's opinion) (Hull 1983, 71). Peter Bowler accurately recognizes Darwin's approval of Owen's results. But he finds it "curious" that Darwin gives credit to Owen "without mentioning that [Darwin's] theory made equal nonsense out of Owen's own explanation" (Bowler 1977, 37). The most widespread misinterpretation of Owen's work, in my view, is its treatment as an "idealistic version of the Argument from Design," mentioned above. To understand the contrast between my interpretation of Owen and the traditional neo-Darwinian interpretation, we must consider the great differences between Owen and Darwin with respect to the explanation of species origins.

7. DARWIN AND OWEN AS PURSUING DISTINCT RESEARCH GOALS

The lack of sympathy shown towards Owen by many neo-Darwinian commentators arises from the great differences between a Darwinian and a structuralist evolutionary theory. In part, but only in part, this comes down to the form-versus-function dichotomy. Neo-Darwinians do not recognize structural explanations as fully explanatory because such explanations tend to ignore natural selection. The constraint-versus-adaptation debates of the 1980s illustrated this contrast (Amundson 1994). This may explain why neo-Darwinians do not see Owen as an important figure. But it does not explain why they misunderstand Darwin's own attitude towards Owen. The contrast between Darwinian theory and structuralist theory is more than the form-versus-function dichotomy.

The structuralist research program grew out of comparative morphology and embryology during the early part of the nineteenth century. The goal was the *understanding of organic form* (Nyhart 1995). The first correspondences to be noticed were between the Natural System of taxonomy and the patterns of embryological development.[9] The focus of attention was not overall patterns of similarity, but the *forms* of organisms. It was assumed that bodily forms indicated relatedness. Embryology was of much more interest to this field than (say) ecological or behavioral studies, because embryology revealed *the origin of form in the individual organism*. A third set of parallels was seen to be the origins of forms in the fossil record (*Limbs,* 56; for an example of this reasoning). As structuralism replaced Bridgewater adaptationism, several authors came to see the structural patterns of Unity of Type as pointers towards a naturalistic explanation of species origins. Owen had written to a publisher in 1848 that he knew of six possible modes of natural species origins. Five of them involved embryological development and some degree of saltation (at least as seen from the Darwinian perspective), and the sixth was gradual change within a species (Rupke 1994, 226). Neither Unity of Type itself, nor such theoretical entities as the Vertebrate Archetype, were intended as ultimate causes.

In my view, the neo-Darwinian commentators have such a bleak view of Owen's contributions to Darwin's thought because they misunderstand the scientific significance of Unity of Type in 1850. They interpret Owen's Unity of Type as a purported ultimate explanation. This is why transcendental anatomy is interpreted as a version of the Argument from Design: the Platonic ideas in God's mind are interpreted as the

---

9. The geometry of the parallel was first seen to be linear, with taxonomy represented by the *scala natura* and embryogenesis seen as a linear progression from simple to complex. This was replaced by von Baer's theory of branching embryogenesis, paralleled to a groups-within-groups taxonomic arrangement.

historically remote "ultimate causes" of the forms of species. But transcendental anatomy was not in the business of ultimate causes, and Darwin recognized that fact. The structuralists of the 1840s, including Owen, were trying to discover phenomenal laws, not ultimate causes. Their ambition was to contribute in the manner of Kepler, not Newton. Ironically, neo-Darwinian commentators have explained very well the basis of this methodological standard (Ruse 1979; Hull 1983). Darwin did not criticize the transcendental anatomists for failing to explain Unity of Type: he congratulated them for discovering it.

Darwin did use Unity of Type as crucial evidence for the fact of evolution. But he did not pursue the structuralist evolutionary theory that Owen had been hinting at in 1849. Darwin convinced the world of the fact of evolution, but natural selection was not accepted as its primary mechanism until the 1930s. The major research program in evolution during the late nineteenth century was "evolutionary morphology," a successor to the morphological program that had flourished for decades on the continent, and had been anglicized by Owen (Nyhart 1995; Bowler 1996). Notable figures in this program were Ernst Haeckel, Carl Gegenbaur, and Frank Balfour. Although Darwin was recognized as the founder of evolutionary biology, Owen's influence was far greater on the actual forms that explanations took during this period (Di Gregorio 1995).

Evolutionary morphology attempted to understand the ways in which changes in embryological processes would lead to evolutionary changes in body forms, and thereby to new species. The program broke down in the 1890s. The processes of embryogenesis proved to be so complex that the next step—understanding how embryogenesis *can change,* thereby causing evolution—looked insurmountable. The Owen-like program of evolutionary morphology would have to be postponed until better embryological knowledge was available. Before that happened, the Evolutionary Synthesis intervened. Neo-Darwinians rejected structuralist evolutionary theories, labeling them as "typological" and perniciously "idealistic." Sixty years later the growth of new molecular knowledge about development led to a rebirth of interest in structuralist biology, in the form of evo-devo. Many evo-devo thinkers recognize the nineteenth century evolutionary morphologists as intellectual precursors (Hall 1999, 2000). Fewer of them recognize Owen as a precursor, but perhaps the republication of *Limbs* will change that.[10]

How does a Darwinian or neo-Darwinian research program differ from this brand of evolutionary morphology? The obvious difference is adaptationism. But equally important is the contrast in explanatory goals between the programs. The structuralist goal has always been to *explain organic form*. In the structuralist program, species are considered as *forms*. Structuralist explanation requires an understanding of the production of form during embryogenesis, and a further understanding of how the embryological process of form-production *can change*. Changes in the process of form-production will create new forms. New forms constitute evolution.

In contrast, Darwin's explanatory goal was never

10. Scott Gilbert, evo-devo practitioner and historian of science, recognized Owen's importance as early as 1980. He ended the paper with a quotation of the passage on species origins from Owen's *Archetype* quoted above, which he said was "as true today as in 1848" (Gilbert 1980, 487). Brian Hall (1994) edited an anthology of articles on homology that was dedicated to the 150th anniversary of Owen's definition of the term.

to *explain form*. Form, for Darwin, was merely one of many characteristics of a species. Darwin was interested in *all* species traits, not just form. Behavior, instincts, emotions . . . name a species characteristic and Darwin (or neo-Darwinians) will try to explain it. Darwinian explanations require an ancestral trait (often hypothesized), heritable variation in that trait, and an environmental cause of differential reproduction. *Darwinian explanation of a trait does not require an understanding of the ontogeny of the trait.* The irrelevance of ontogeny to Darwinian explanation marks the crucial difference between Darwinian and structuralist evolutionary theories.

Structuralists want to explain form *qua* form, not form *qua* heritable character. Owen's first comments about species origins in *Archetype* referred to "the successive introduction of specific forms [species-forms] of living beings." Structuralists *must* explain form in terms of the form-generating processes of embryogenesis. They cannot skip (like Darwin does) from phenotype to phenotype, explaining change by selection on phenotypes alone. They must incorporate the ontogeny of a character, and the ways in which that ontogeny can change, in order to produce a structuralist explanation of the character's evolution. Rudy Raff, prominent evo-devo researcher, explained the difference between neo-Darwinism and evo-devo in this way: "They're interested in species. We're interested in bodies" (Amundson 2003, 253).

Owen was a structuralist. Working towards the natural causes of species origins (the origins of *new forms*) was surely a motivating factor in *Limbs* and other writings of the 1840s. Darwin, having already devised a very different account of species origins, did not even recognize Owen's evolutionary ambition. Nevertheless, Owen's work paid off. Darwin used his insights to great effect in the *Origin,* and they contributed to evolutionary morphology. Evo-devo researchers are coming to recognize their historical predecessors in the evolutionary morphologists and perhaps Owen. Nevertheless, the task of accommodating the very different insights of Darwin and Owen is still before us.

REFERENCES

Amundson, Ron. 1994. Two concepts of constraint: Adaptationism and the challenge from developmental biology. *Philosophy of Science* 61:556–78.

———. 1996. Historical development of the concept of adaptation. In *Adaptation,* ed. Michael Rose and George V. Lauder, 11–53. New York: Academic Press.

———. 2005. *The changing role of the embryo in evolutionary thought: Roots of evo-devo.* Cambridge: Cambridge University Press.

Appel, Toby A. 1987. *The Cuvier-Geoffroy debate: French biology in the decades before Darwin.* New York: Oxford University Press.

Barrett, Paul H., Peter J. Gautrey, Sandra Herbert, David Kohn, and Sydney Smith. 1987. *Charles Darwin's notebooks, 1836–1844.* Ithaca, NY: Cornell University Press.

Bell, Sir Charles. 1833. *The hand: Its mechanism and vital endowments as evincing design.* London: William Pickering.

Bowler, Peter J. 1977. Darwinism and the argument from design: suggestions for a reevaluation. *Journal of the History of Biology* 10:29–43.

———. 1996. *Life's splendid drama.* Chicago: University of Chicago Press.

Camardi, Giovanni. 2001. Richard Owen, morphology, and evolution. *Journal of the History of Biology* 34:481–515.

Chambers, Robert. 1844. *Vestiges of the natural history of creation.* London: Churchill.

Corsi, Pietro. 1988. *Science and religion: Baden Powell and the Anglican debate, 1800–1860.* Cambridge: Cambridge University Press.

Darwin, Charles. 1859. *On the origin of species.* London: John Murray.

———. 1909. *The foundations of the origin of species: Two essays written in 1842 and 1844 by Charles Darwin.* Cambridge: Cambridge University Press.

Desmond, Adrian. 1989. *The politics of evolution.* Chicago: University of Chicago Press.

Di Gregorio, Mario A. 1995. A wolf in sheep's clothing: Carl Gegenbaur, Ernst Haeckel, the vertebral theory of the skull, and the survival of Richard Owen. *Journal of the History of Biology* 28:247–80.

Gerhardt, John, and Marc Kirschner. 1997. *Cells, embryos, and evolution.* Malden: Blackwell.

Gilbert, Scott F. 1980. Owen's vertebral archetype and evolutionary genetics—a Platonic appreciation. *Perspectives in Biology and Medicine* 23:475–88.

Gould, Stephen Jay. 2002. *The structure of evolutionary theory.* Cambridge: Harvard University Press.

Hall, Brian K. 1994. *Homology: The hierarchical basis of comparative biology.* San Diego: Academic Press.

———. 1999. *Evolutionary developmental biology.* 2nd ed. Dordrecht: Kluwer.

———. 2000. Balfour, Garstang and de Beer: The first century of evolutionary embryology. *American Zoologist* 40:718–28.

Hull, David L. 1973. *Darwin and his critics.* Chicago: University of Chicago Press.

———. 1983. Darwin and the nature of science. In *Evolution from molecules to men,* ed. D. S. Bendall, 63–80. Cambridge: Cambridge University Press.

Lenoir, Timothy. 1982. *The strategy of life.* Chicago: University of Chicago Press.

Mayr, Ernst. 1961. Cause and effect in biology. *Science* 134:1501–6.

———. 1982. *The growth of biological thought.* Cambridge: Harvard University Press.

Müller-Wille, Staffan. 1995. Linnaeus concept of a "symmetry of all parts." *Jahrbuch für Geschichte und Theorie der Biologie* 2:41–47.

Nyhart, Lynn. 1995. *Biology takes form.* Chicago: University of Chicago Press.

Oken, Lorenz. 1807. *Über die Bedeutrung der Schadelknochen.* Jena: Göpferdt.

Ospovat, Dov. 1981. *The development of Darwin's theory.* Cambridge: Cambridge University Press.

Owen, Richard. 1843. *Lectures on the comparative anatomy and physiology of the invertebrate animals.* London: Longman Brown Green and Longmans.

———. 1848. *The archetype and homologies of the vertebrate skeleton.* London: J. van Voorst.

Padian, Kevin. 1995. Form versus function: The evolution of a dialectic. In *Functional morphology in vertebrate paleontology,* ed. Jeff Thomason. Cambridge: Cambridge University Press.

Raff, Rudolf A. 1996. *The shape of life.* Chicago: University of Chicago Press.

Richards, Eveleen. 1987. A question of property rights: Richard Owen's evolutionism reassessed. *British Journal for the History of Science* 20:129–71.

———. 1994. A political anatomy of monsters, hopeful and otherwise. *Isis* 85:377–411.

Richards, Robert J. 2002. *The Romantic conception of life.* Chicago: University of Chicago Press.

Roget, Peter Mark. 1834. *Animal and vegetable physiology considered with respect to natural theology.* 2 vols. London: William Pickering.

Rupke, Nicolaas A. 1993. Richard Owen's vertebrate archetype. *Isis* 84:231–51.

———. 1994. *Richard Owen: Victorian naturalist.* New Haven: Yale University Press.

Ruse, Michael. 1979. *The Darwinian revolution.* Chicago: University of Chicago Press.

Russell, Edwin S. 1916. *Form and function.* London: John Murray.

Sloan, Phillip R. 1992. Introduction: On the edge of evolution. In *The Hunterian lectures in Comparative Anatomy,*

May and June 1837, ed. Richard Owen, 3–72. Chicago: University of Chicago Press.

————. 2003. Whewell's philosophy of discovery and the archetype of the vertebrate skeleton. *Annals of Science* 60:39–61.

Winsor, Mary P. 2003. Non-essentialist methods in pre-Darwinian taxonomy. *Biology and Philosophy* 18:387–400.

RICHARD OWEN'S QUADROPHENIA

THE PULL OF OPPOSING FORCES
IN VICTORIAN COSMOGONY

Kevin Padian

The world of Victorian biology—the word, coined
by Lamarck decades before the queen's reign, had a
very different compass in Victorian Britain, let alone
today—was dauntingly complex. It was a world that
was moving out of the Enlightenment and into the
Industrial Age (Wilson 2002), where science, engi-
neering, industry, and ambition were undoing many
traditions of class structure, economics, politics, and
culture (Fowles 1969). The very roots of biological philosophy were being transplanted from an orderly, theo-
centric pot to a "tangled bank" (in Charles Darwin's
words) of material causes and open outcomes. A reading of Darwin's *Origin of Species* uncovers arguments
about form, change, embryology, and ecology that have
no close counterpart today. In quite a different way,
the allusions to divine plan that are standard in writ-
ers from the early Enlightenment through William
Buckland and Gideon Mantell to Richard Owen himself often strike today's readers as quaint, misguided,
or philosophically outdated. And the suppression of
works and the ruination of careers such as Mantell's
(Desmond 1982; Dean 1999; Cadbury 2000) and Rob-

ert Edmond Grant's (Desmond 1989), that had an ideological basis in class and station as well as personal
and scientific qualities, are based in specific contemporary issues of politics and religion that are completely
foreign to us today.

In this introductory essay, I want to provide a brief
and general overview to the philosophy of morphology
that Owen describes in *The Nature of Limbs,* particularly insofar as it relates to what must be regarded
as its companion volume, *Report on the Archetype and
Homologies of the Vertebrate Skeleton* (1848). Works by
Desmond (e.g., 1979, 1982, 1989), Rupke (1985, 1993,
1995), Gruber and Thackray (1992), and Sloan (1992)
should be consulted for a fuller account of Owen's philosophical influences, and particularly Desmond (1989)
for the sociopolitical context of morphological and evolutionary thought in the 1820s and 1830s. I want to
outline briefly what Owen was trying to do in both
Archetype and Homologies and *The Nature of Limbs,*
and why these two works were so central to his philosophy. And I want to contrast some of the currents of
scientific thought that Owen had to draw from, respond
to, support, and oppose in order to establish his worldview as central to his times. The stakes were high:
although at one time or another several British biolo-
gists received the sobriquet of "the English Cuvier,"
Owen was the undisputed heir, if not the philosophical
descendant, of the late baron, and coveted his primacy
in comparative anatomy. Owen was expected to stem
the tide of materialistic, atheistic morphological and
evolutionary thought. And only by doing this successfully would he realize his dream of founding a natural
history museum like Cuvier's.

OWEN'S PHILOSOPHICAL CORNERSTONES

What I have called Owen's philosophical cornerstones (Padian 1995a, 1997) were not original to him; but as Russell (1916) pointed out, Owen was a great synthesizer who took what was useful and discarded what was not from the concepts of previous authors, often of quite different philosophical schools. These concepts were the Archetype, the Vertebral Theory, and Homology and Analogy (fig. 0.1). At the time that he gave the Hunterian Lectures on comparative anatomy as a young assistant conservator in the Royal College of Surgeons in 1837 (Sloan 1992), Owen accepted the Vertebral Theory, but he was still using homology and analogy somewhat interchangeably, as it was then being done in France; and he clearly did not accept the concept of Unity of Type that was necessary for the model of the vertebrate Archetype (Padian 1995a). By 1843 he had straightened out homology and analogy, and the three cornerstones had been collectively laid by 1846, when he gave his long lecture entitled "Report on the Archetype and Homologies of the Vertebrate Skeleton" to the British Association for the Advancement of Science. That lecture appeared in book form in 1848, and in 1849 he lectured to the Royal Institution on "The Nature of Limbs," which was published later that year.

To begin with the most familiar concept today, Owen (1843) is responsible for the use of the terms homology and analogy as we now use them—with the important qualification that he did not accept that common descent was the underpinning of homology (Darwin's theory established this: Ellegård 1958; Glick 1988). For Owen, homologous organs were "the same organ in different animals under every variety of form and

Figure 0.1.

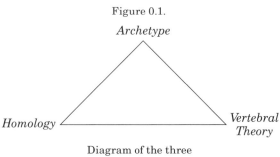

Diagram of the three
cornerstones of Owen's thought.

function." Homologues were distinguished from ana-
logues, which were "a part or organ in one animal
which has the same function as another part or organ
in a different animal." Thus, homologous organs were
possessed by creatures that shared a common plan;
analogous organs were not built on such a plan. So, for
example, the hands of the mole, bat, horse, and other
animals that Owen illustrates in *The Nature of Limbs*
are homologous, even though they have very different
functions. In contrast, the bones of the bird and bat
wingspars are not formed in just the same way, nor are
their airfoils made of the same kinds of tissues, so they
are not homologous but analogous in those respects.

Owen relied on three criteria of homology. Most
important of these was *position:* an organ or tissue
would be consistently adjacent to another in various
organisms built on the same plan. A second criterion
was *histology:* the organ needed to be made of the same
kind of tissue in different organisms in order to be
comparable and therefore homologous. Finally, there
was the criterion of *ontogeny:* did the tissues in question develop in the same way, from the same embryological precursors? Working among these criteria,

Owen was able to develop many examples of homology that contradicted hypotheses of previous workers (for example, Oken's idea [*Limbs,* 41 and 81] that the limbs represented free ribs). Owen knew, and explained, that in the course of the progression of forms through time (he is careful not to use the word "evolution" but he accepted, as any rational person of the time had to, the superposition of geologic strata and the evolutionary parade of their fossil forms), any of these criteria could be altered, covered, or eradicated by locally adaptive requirements. But if one looked carefully enough and worked among criteria, the underlying homological plan would be revealed. This is why, for example, he was able to assert that the forelimbs are always correlated with the fourth vertebra, whereas the condition could not be standardized for the hindlimbs (see below).

Darwin's evolutionary view of homology did not change the criteria; it just provided a deeper explanation for homology, one that supplanted Owen's archetypal rationale and replaced it with common descent. The mistake is often made in textbooks that homology is evidence for evolution, and this is seized upon by advocates of Intelligent Design and other creationists (e.g., Wells 2000). Owen would have been as appalled as we at the misuse. Evolution, in the form of common descent, explains the homological resemblances manifested in the pre-Darwinian criteria that Owen advanced. These criteria are, in a sense, the *diagnosis* of homology, whereas common descent *defines* it (Padian 1995a, 1995b, 1997). In a similar way, in the phylogenetic system, taxa are *defined* on the basis of common ancestry, but *diagnosed* on the basis of shared common features (Rowe 1987).

Owen distinguished three kinds of homology. *Spe-*

cial homology allows us to establish that structures in different organisms correspond to each other; so, for example, the humeri of a mole, a bat, a whale, and a human correspond, despite all of their adaptive modifications of form. *General homology* is a more theoretical concept: it "stands to" the "fundamental type," as Owen put it, by which he meant that it corresponded to a particular structure in the Archetype (see below). But, as Amundson points out in his essay in this book, citing Camardi (2001), the concept of general homology relates fundamentally to a corresponding structure in the ideal vertebra, the plan on which all vertebrates are built—providing, of course, that one is discussing vertebrates, which Owen always was. (It is an unanswered question whether he would have accepted an Archetype for plants, based on Goethe's theory; and whether he would have accepted one or more Archetypes for invertebrate forms—and if so, whether he would use Cuvier's scheme of four *embranchements* or one of the alternative schemes proposed by Lamarck, Geoffroy, d'Orbigny, and others.) Finally, serial homology applies to repeated structures along, in the case of vertebrates, an anteroposterior axis. So, for example, the vertebrae are all homologous to each other; and, as we shall see, the limbs are homologous not only to each other but to parts of the vertebrae. And so are the skull bones.

The concept of homology was legitimized, in Owen's worldview, by the Vertebral Theory, the idea that the skull was simply a series of fused vertebrae (the number was commonly recognized as between three and six; Owen was committed to four). It may seem strange to us to think of the skull as fused vertebrae, but as he points out in *Limbs* (55), if you want to find homologies, you have to look at the simplest forms, not the most complex; these have undergone the least modification. This is why he used the fishes so extensively to establish homologies (see below). Owen used his criteria of position, histology, and ontogeny to show what specific parts of the ideal vertebra could be modified to produce all sorts of organs.

Owen explained his use of the Vertebral Theory in *Archetypes and Homologies,* from which the accompanying figures are taken (Russell [1916] explained much of this and more, but the reprint of his classic book is now, sadly, also out of print). The ideal vertebra (fig. 0.2a) contains all the formal elements that correspond to every kind of skeletal bone found among the vertebrates. There is both a dorsoventral and a dextrosinistral symmetry to its parts. The neural and haemal arches house the nerve tube and the digestive tube, respectively, above and below the centrum. Extensions of these arches, adaptively modified, can become nearly any structure in the skeleton. So, for example, Owen uses a segment through a bird's thorax (fig. 0.2b) to demonstrate that even the most elaborate structures are all part of the ideal vertebra, however modified. Notice, for example, that the "h" in the diagram of the bird thorax—ostensibly the keel of the sternum, which is actually not a separate bone in birds—is meant to represent the haemal spine of the ideal vertebra. The other parts can be traced accordingly.

It is strange to think of "cranial vertebrae," but as early as 1837 in his Hunterian Lectures Owen was identifying the bones of the skull in these terms. The four cranial vertebrae can be seen clearly in his exploded view of a cod skull (fig. 0.3). To realize his scheme, Owen had to work out the special homologies of all vertebrate skull bones, despite their losses, fusions, modifications, and differences in names given

Figure 0.2a.

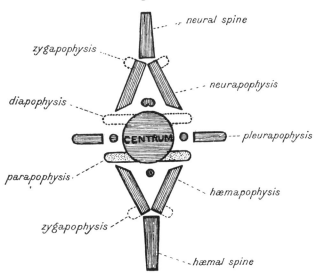

Owen's conception of the ideal vertebral segment. Compare
with figure 0.2b, which is an actual segment through a bird's
thorax. Note how the names of vertebral structures in figure 0.2a
correspond with the abbreviations in figure 0.2b. From Owen
(1848); see *Limbs,* 43. After Russell (1916).

to them in different taxa by different workers. The
establishment of these homologies was in itself a major
advance. The result of his labors, the end of which was
the identification of general homologies with respect to
the vertebrate Archetype, is summarized in the table
in figure 0.4. The first column of the table lists the
parts of the ideal vertebra, and the remaining columns
are headed by the names of the cranial vertebrae;
from front to back they are the nasal, frontal, parietal, and occipital vertebrae. So it can be seen that the
dermal bones of the cranium along the midline of the

Figure 0.2b.

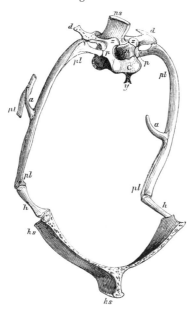

An actual vertebral segment. The letters are abbreviations
of the names of vertebral structures as given in figure 0.2a.
From Owen (1848); see *Limbs,* 42. After Russell (1916).

skull correspond to the neural spines (though they are
paired, whereas the spine is not); and the bones of the
braincase correspond to the parts of the neural arch
that surround the spinal cord in a typical vertebra.
(In his 1837 lectures Owen noted that "the cranium
[is] a dilated continuation of the same central axis,"
by which axis he meant the spinal column [Padian
1995a].) In this scheme, the lower bones of the occipital
vertebrae correspond to the bones of the shoulder girdle and limb. Owen was able to determine this because
in a typical actinopterygian fish the shoulder girdle is

Figure 0.3.

Owen's (1848) use of the cod skull to demonstrate the
correspondence of bones to each of the four "cranial vertebrae"
that he interpreted in the skull. From Russell (1916).

connected to the skull, so by the principle of connections Owen could identify to which parts of the ideal
vertebra the girdle and its appendage belonged.

The Archetype itself (fig. 0.5) looks rather like a bony
amphioxus (*Branchiostoma* had just been described
in 1844), right down to the impression of a ciliated or
villi-ringed oral opening. It became the rationale for
why the Vertebral Theory would work at all, and hence
how one could establish homologies in the first place.
In fact, Owen needed the Archetype to establish the
plan on which the modification of vertebrae in all their
forms was based. He did not realize this early in his
career; preoccupied with the details of anatomical correspondence, he appears not to have worried too much
about Idealistic philosophy. Although he accepted the

Figure 0.4.

Cranial Vertebræ.[1]   (After Owen, 1848, p. 165.)

Vertebræ.	Occipital.	Parietal.	Frontal.	Nasal.
Centra.	Basioccipital.	Basisphenoid.	Presphenoid.	Vomer.
Neurapophyses.	Exoccipital.	Alisphenoid.	Orbitosphenoid.	Prefrontal.
Neural Spines.	Supraoccipital.	Parietal.	Frontal.	Nasal.
Parapophyses.	Paroccipital.	Mastoid.	Postfrontal.	None.
Pleurapophyses.	Scapular.	Stylohyal.	Tympanic.	Palatal.
Hæmapophyses.	Coracoid.	Ceratohyal.	Articular.	Maxillary.
Hæmal Spines.	Episternum.	Basihyal.	Dentary.	Premaxillary.
Diverging Appendage.	Fore-limb or Fin.	Branchiostegals.	Operculum.	Pterygoid and Zygoma.

Owen's (1848) table of homologies of the bones of the
"cranial vertebrae," again from Russell (1916).

Vertebral Theory at least by 1837, his works of the later 1840s are the first explicit statements of the centrality of the Archetype.

The Archetype was supposed to be generalized enough to be able to develop into nearly any kind of form, and its repeated organs are readily homologized both with each other and with any modified forms among the vertebrates. Owen's pattern-coded diagram makes the serial and general homologies plain, and the special and general homologies of the various vertebrate limbs are revealed in plate 1 of *Limbs.*

Figure 0.5.

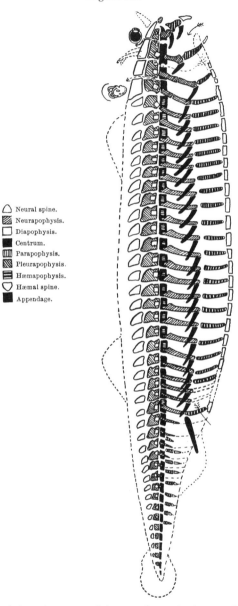

Owen's (1848) concept of the vertebrate Archetype. His
coded legend shows to what parts of the ideal vertebra each part of each skeletal element corresponds. See
Limbs, plate 1, figure 1. Modified from Russell (1916).

In this way, Owen's worldview can be seen as an interdependence of three separate concepts (fig. 0.1). The correspondence of parts among different organisms, and in the same organism, provides patterns that can be described as various kinds of homology. The Vertebral Theory, in some respects, is a case of serial and general homology. But it would not make sense all together if there were not a common plan, an Archetype. Owen may have been reluctant to enunciate a common plan for vertebrates in the late 1830s, so soon after the deaths of Cuvier and Geoffroy, the latter of whom had argued unsuccessfully for this unity across invertebrates (fig. 0.6). Although Owen was gun-shy then about unity of plan, he warmed to the concept considerably within the next decade. On page 70 of *Limbs,* he asserts that his scheme is "no mere Transcendental dream," and at the end of his text he is positively eloquent about the importance of the insights that the Archetype will bring us: it will give humans knowledge of their place in the whole of nature, the unity of type, and "a better conception of their own origin and Creator."

THE OUTLINE OF *ARCHETYPES* AND *LIMBS*

Archetypes and Homologies of the Vertebrate Skeleton is a necessary precursor to *The Nature of Limbs;* without it, the rationale for *Limbs* is only dimly perceived, and Owen's opening arguments about *Bedeutung,* the "meaning" of limbs, is strangely isolated from any philosophical context. The foregoing section has provided a very brief introduction to that context, and so the organization of the two works now may be shortly outlined.

Archetypes and Homologies is basically the seat of

Figure 0.6.

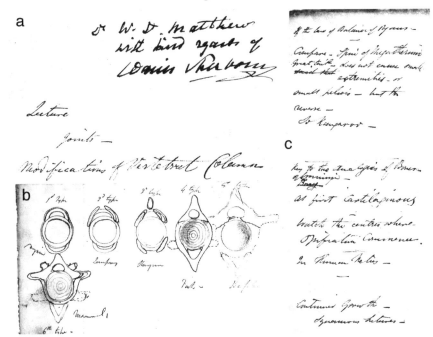

Parts of a "missing" lecture (Padian 1995a) from Owen's 1837
Hunterian series in comparative anatomy (Sloan 1992). *a,* detail of
title page, showing its focus on the vertebral axis (the dedication
is from Charles Davies Sherborn, antiquarian, bibliographer, and
amanuensis to Owen at the British Museum of Natural History;
the recipient is William Diller Matthew, vertebrate paleontologist
for many years at the American Museum of Natural History
in New York and, from 1927 until his death in 1930, at the
University of California, Berkeley). *b,* a set of Owen's drawings
of various stages in vertebrae, from hagfish to mammal, on a slip
of paper associated with this manuscript. *c,* another slip of paper
associated with the manuscript. Owen seems to be puzzling out
problems of homology, conditions of existence, and embryology.
He is considering the small pelvis of the giant ground sloth
Megalonyx, and notes that it is not associated with small limbs
and tail, but rather the reverse (as with the kangaroo). He next
makes a note to consider relative position of elements in the most
basic forms; and finally remarks on the "ingenious" sutures of
the fetal human skull. For details see Padian (1995a, 1997).

explanation of Owen's three cornerstones. He begins by justifying the need to standardize terms for common anatomical structures, to move beyond mere descriptive phrases to a general signification. He regards them as homologues not simply because they develop in the same way (as Geoffroy said, mistakenly quoting, as Owen thought, the German morphologists). The mode of development may change: "thus the 'pars occipitalis stricte sic dicta' &c. of Soemmerring is the special homologue of the supraoccipital bone of the cod, although it is developed out of pre-existing cartilage in the fish and out of aponeurotic membrane in the human subject. . . . The femur of the cow is not the less homologous with the femur of the crocodile, because in the one it is developed from four separate ossific centres, and the other from only one such centre" (Owen 1848, 5). Owen explains the three kinds of homology by stating that special homology is based on the criterion of position with respect to other structures, general homologies are those that "stand to the fundamental or general type," and serial homologies refer to repeated structures that have corresponding parts, which he calls *homotypes.*

Owen then goes about standardizing the terminology of skull bones, with profuse respects paid to the memories of those great anatomists whose terms he is frequently forced to replace, but "only where nature seemed clearly to refuse her sanction to them" (ibid., 9). His justification is not only clarity but also the correspondence of structures to parts of the ideal vertebra, and he is explicit in faulting the work of previous authors (Geoffroy, for example) because they only considered special and not general homology. He proceeds to spend the next fifty pages discussing specific examples of special homology in the skull, moving deftly among a tremendous variety of features and taxa, with an intimate understanding of development, connections, variation within groups, and anomalies. This is where Owen becomes the consummate lab rat, the morphology freak, the historian and true philosopher of biology. He seems to forget himself temporarily; he loses himself in the wonder of comparisons, the diversity of adaptive modification, the frankly admitted puzzles of variation, and his prose soars as a result, becoming far less didactic and pompous than usual. These pages are beautifully written, a whirlwind tour de force of comparative anatomy, all the more so because the overarching motive of the description is to establish special homologies so that the reader is fully prepared for the justification of general homology that will follow.

Owen then turns to general homology; in parallel to the previous section, he begins with the history of the subject. His first task is to dispose of Cuvier's predilections, always a delicate task, and all the more so here because Cuvier did not concern himself with general homology. However, because Cuvier is the 500-pound gorilla in the roomful of anatomists, Owen is at pains to show that Cuvier's ideas on special homology, which were so entailed with functional ends, can be only a poor guide to the true general homologies of skeletal structures. Clearly teleological (purely functional) explanations will not work here, he says, and if with Cuvier we must disavow the notion that form is simply "some higher type of organic conformity" known only to the Creator, then the only possible explanation that remains is chance, which we must obviously reject. (Owen used exactly the same rhetoric in *The Nature of Limbs,* 40.)

Owen, with almost uncharacteristic generosity, then produces a passage that gives Lorenz Oken full and unstinting credit for establishing the concept of general homology. Much like the (evidently apocryphal) story of Goethe with the sheep skull (see below), Owen quotes Oken's epiphany in the Hartz Mountains when he stumbled over a deer skull and realized that it was composed of a series of vertebrae. Oken became the first to work out the detailed homologies, both serial and general, of the skeletal bones. But here we see why Owen was able to be so generous to Oken about this; he alludes to a point farther along in his own work where he will dispense with Oken's notion that the limbs, by homological relations, can be none other than free ribs.

To explain general homology, which becomes in Owen's hands a frankly transcendental concept, he first describes a "typical" vertebra, and shows its segmental manifestation in a bird (figs. 0.2a and 0.2b) and in a fish. As he did with the skull vertebrae, he explains how its parts are positionally correlated, and how they develop ontogenetically. He then discusses the form of the trunk vertebrae and their modifications (including the hips), and then proceeds to show how the skull is a series of fused vertebrae. Again he must be careful here, because Cuvier rejected the proposition except insofar as certain structures, like the occipital bone, have the same functions of movement as the ends of adjacent cervical vertebrae, and so resemblances could be expected. (Owen also has to be careful because Cuvier's authority was very well respected in Britain, and because he needed to invoke Cuvier's strict empirical approach in countering Geoffroy's transcendental excesses elsewhere.) Owen's success in defeating this view depends on his ability to show that Cuvier was either confused or inconsistent about his understanding of the general homology of cranial structures, and on the supposition that had Cuvier apprehended the true relations of their parts, he would have had little choice but to accept their higher homological relations, regardless of similarities or differences in function.

Owen concludes the main body of his text with a discussion of serial homology, which is mostly historical and can be brief because he has already made all of the points necessary to the acceptance of this principle by his audience. He finishes with a very long explanation of the various illustrations of his plates.

The Nature of Limbs is very similar in structure and argumentation to *Archetypes and Homologies;* it is briefer, and in general more accessibly written because it does not demand the extent of knowledge of comparative anatomy as its predecessor. It is more didactic and oratorical in style, reflecting its origin as a lecture more clearly than many of Owen's published treatises. As Sloan (1993) has shown so well for Owen's Hunterian Lectures, the audience in question was likely to comprise some medical students (many of whom were befuddled by his abstruse philosophizing and terminology) but also many members of his largely Tory supporters, who came more for social reasons than for scientific ones (see also Desmond 1982, 1989). In the 1840s the anonymously published *Vestiges of the Natural History of Creation* had created considerable public stir about evolution and transmutation, and Owen was expected to provide a response to these issues in terms of pure and applied morphology—a circumstance that put him in a delicate position (Richards 1987; Secord 2000, 421–25). How would Owen explain the resemblance of plan and the diversity of function of vertebrate limbs? Would he espouse Cuvier's functionalism or Geoffroy's formalism? Would his teleology restrict itself to the pure functionalism of Cuvier or extend to the divine design arguments of the natural theologians? How would he manage to dance around the question of Unity of Plan?

Limbs begins with a discussion of the German word *Bedeutung,* the closest approximation to which in English is the term "signification" or "meaning." Immediately Owen's debt to German transcendental morphology is apparent. What "meaning" can limbs have, apart from some sort of metaphysical construct? But this, in fact, is just what Owen is trying to establish. He cannot lead into it too quickly, however. First, he describes special and general homology. Special homology leads him to a discussion of comparability of function. Limbs can be adapted to many purposes, Owen says, but function is no guide to commonality of plan. Because parts of limbs may disappear from their final use, we know that they are built on a plan of organization, a primary *unity of type.* This is a critical clue that Owen has made his peace with any misgivings he may once have had about Geoffroy's theory of unity of type.

Owen then moves to serial homology, choosing a difficult example: the scapular arch. There is variation in the parts of the shoulder, Owen admits, but the pectoral girdle is serially homologous to the pelvis. Although there are some difficulties in establishing the homologies of several foot bones, the principle of connections is the most reliable guide to identifications. He can depend on this, he notes (*Limbs,* 38–40), because there is a "prescient operation of One Cause of all organization." That "one cause" is ultimately the Creator, but Owen equally has in mind its manifestation in the secondary phenomenon that he has introduced as the Archetype.

CAN ARCHETYPES EVOLVE?

And so, Owen asks, what is finally the nature of limbs?

Cuvier was mute on the question, Oken regarded the limbs as free ribs, and Carus had what appeared to Owen a vague and imprecise formulation of the limbs as modified vertebrae. Owen regards them as a "natural group of bones," the "primary division of the endoskeleton," despite the varied "adaptive mask" that their functions in life compel them to wear. By recurring to his diagrams of the ideal vertebra and the thoracic vertebra of a bird (here, figs. 0.2a and 0.2b), Owen can identify the elements that correspond to the bones of the girdles and limbs. The general homology of the scapular arches are of primary concern (ibid., 46ff.). For Owen, in terms of general homology the coracoid is the haemapophysis of the scapular arch, which he shows through the principle of connections. This is where he stresses the need to examine primitive vertebrates, in which the "original" archetypal relationships are far less modified. In more progressive ("higher") forms, these bones have been displaced from their primitive position "for a special purpose" (ibid., 50). He is thus able to argue, through the principle of connections, that the hands and arms are actually parts of the head (again, the fourth or occipital cranial vertebra). Through serial homology, the ilium, ischium, and pubis of the pelvic girdle correspond to the scapula, coracoid, and clavicle of the shoulder girdle. And these, in turn, correspond generally to parts of the ideal vertebral segment.

The final pages of *Limbs* become more philosophically reflective. There is a strange passage (*Limbs,* 83) in which Owen states (in very careful language) that the Archetype would be expected to be found on other planets that were habitable; he put a similar passage on pages 102–3 of *Archetypes and Homologies*. This is no speculation on interstellar travel, but an affirma-

tion of the universality of the secondary mode of operation that produces the Archetype and its derivatives,
the diversity of life. The question of the possibility of
a plurality of worlds has a long history reaching back
to the ancient Greeks, and was mooted in Christian
tradition many times since the medieval period (Crowe
1986; Brooke 1991). On one hand, if God were omnipotent, He could have created any number of worlds, but
it was dangerous to imply that he had. On the other
hand, if Christ had been sent to redeem humanity by
his death, it was considered ludicrous and heretical by
Luther and others that he could have done so repeat-
edly on all possible worlds of an infinite universe.

This debate was relevant to Owen's philosophy in
part because it was a preoccupation of his friend William Whewell, a fellow Lancastrian ten years Owen's
senior, who was master of Trinity College, Cambridge,
and the greatest Victorian exponent of the history
and philosophy of the inductive sciences. Whewell was
(until 1853, when he reversed himself) a proponent of
the plurality of worlds. He broached the question in his
volume in the series of Bridgewater Treatises, *Astron-
omy and General Physics Considered with Reference
to Natural Theology* (1833), repeating the commonly
accepted assumption that many planets in our solar
system and presumably others would be habitable. As
Crowe (1986, 270) wryly notes:

> Whewell states: "with regard to the physical world, we
> can at least go so far as this;—we can perceive that
> events are brought about, not by insulated interposi-
> tions of divine power exerted in each particular case,
> but by the establishment of general laws" (Whewell
> 1833, 267). This statement is especially significant
> because Whewell was later to encounter it, probably
> to his distress, as the lead quotation in the book that
> did the most damage to the cause (i.e., the promotion

of Natural Theology) so fully financed by the Earl of Bridgewater. This book was Charles Darwin's *The Origin of Species.*

Owen's reference to the possibility of life on other worlds was not idle speculation, then, but a scientific contribution to an issue with great theological resonance for his audience and for the philosophical considerations of his time. If the Creator suffered life to exist on other planets, Owen asserts, such beings would presumably respond to light with dioptic organs much like those of animals on Earth; and if they moved on the basis of a vertebrate organization, there can be no doubt that there would be variations the like of which never existed on Earth, but that would be discernible and predictable to those who truly understood the Archetype. This lawlike operation of secondary causes, bizarre as it must have sounded to many in his audience, was intended to ensure that Owen could bring biology into the same realm of predictability that physics and chemistry enjoy. In the same spirit, he attacks the notion that "nothing is made in vain," the hyperadaptive claim of the Design advocates; nature works according to a Plan, but the Plan is one of organization, not of optimality of function.

Finally, he asserts, knowledge of the Archetype will give us a better sense of our place in the world and of the Plan of the Creator. This passage could be construed as a sop to the Paleyites, but Owen has just finished rejecting one of their favorite ploys. However, he quotes Cudworth on the role of investigation of nature in illuminating theology. For Cudworth, the atheists say that if a Deity created the world, then the "Idea" of the world must precede it; so Knowledge would come before Things. But atheists do not accept that pre-existent mind, so there cannot be an Archetype on which natural forms are based. This reasoning allows Owen to step in and assure us that because we know that there is an Archetype, the knowledge of Man must have preceded Man, and so atheism is impossible: "the divine mind which planned the Archetype also foreknew all its modifications." Owen is able to assert the existence of a Creator from his works, but in a very different way than Paley and the natural theologians did (see below).

Owen ends by admitting that we do not know the secondary causes of morphology—that is, the laws by which shapes are produced and modified. But if we can call these causes "Nature," then we can see how "such organic phaenomena" (i.e., animal forms) could have progressed through time. As Rupke (1993) showed, this relatively straightforward admission by Owen that form could change by natural means brought a rain of criticism, from old Adam Sedgwick to the equally retrenchant literary critics in popular periodicals. Special creation was a difficult vampire to lay to rest, especially along the Oxbridge axis of Paleyite sentiment. Owen was obliged to restate, not without some authoritative asperity, his faith in Divine creation without giving away his right to search for natural mechanisms—using some verbal *legerdemain* that did not entirely mollify his critics (Rupke 1993).

Evelleen Richards (1987) deftly showed how Owen's treatment of the evolution question was complicated by Robert Chambers's (1844; anonymously published) *Vestiges of the Natural History of Creation,* building on earlier analyses by Desmond, Ruse, and Ospovat. Owen accepted evolution but balked at transmutation because it was difficult to propose a mechanism that was not purely naturalistic (materialistic) but also did not rely too explicitly on divine intervention and design. Despite imprecations from his supporters,

Owen (like Whewell) chose not to publish an extensive, specific critique of *Vestiges* (as he did in 1860 of Darwin's *Origin* in the *Edinburgh Review*). Instead, he wrote a long and deferential letter to the author, pointing out misinterpretations from his own works and clues to evolutionary mechanisms that he hoped the author would consider in planning future editions. But Owen did not escape rebuke for his failure to slam the *Vestiges*. Adam Sedgwick worried that Owen was on a slippery slope of progressive development, and the Puseyites (conservative Catholics of the day) excoriated him for abandoning the doctrine of creation for a science that was incompatible with it (Richards 1987). In 1849, following the publication of the *Nature of Limbs,* Owen was attacked in the weekly *Manchester Spectator* for what the anonymous writer deemed his "scientific Pantheism." This sin was shared by the author of *Vestiges* and the German transcendentalist Lorenz Oken, from whom Owen had borrowed heavily (and by whose work, *Lehrbuch der Naturphilosophie,* Owen had been embarrassed two years before when he recommended its translation and publication by the Ray Society, a project that raised alarms for its "symptoms of unsound religious principle" [Richards 1987]). Owen argued that his intent in *Limbs* "was not to support the development hypothesis, but rather to argue that his morphology demonstrated the pre-existence of the archetypal ideal in the mind of the Creator, and thus refuted the argument of the old Pantheists and atheists that mind did not precede matter" (Richards 1987, 165). This kind of public (yet anonymous) confrontation is exactly what Owen did not want, because it put him in a no-win situation: no answer would have satisfied all his supporters and critics at once, and he needed the freedom of ambiguity.

THE ARCHETYPE:
PLATONIC, ARISTOTELIAN, OR NEITHER?

Rupke (1993, 1995) has shown that Owen essentially got his concept of the Archetype from Carus, though without attribution. As we have seen, when Owen cited his intellectual forebears it was generally either to pander to them, as to Cuvier, especially when their conclusions were not particularly germane to his, or to vilify them if they got too close to his ideas and in his view got them wrong. As Rupke notes, Owen did not see that others paved the way for him; rather, they committed basic faults that he was at pains to rectify, so there was no real need to cite them extensively. He felt justified in using ideas that were already in the literature, because in his view less credit should be given to the first person to propose a concept than to the one who got it right.

There has been some discussion in the literature about how "real" the Archetype was to Owen, and what it meant as a concept. Desmond (1982, 1989) has expressed the view of most commentators that the Archetype was a Platonic concept that eventually became more tangible in Owen's later writings. Rupke (1993, 1995) has argued that the Archetype was not Platonic but Aristotelian, because it was a model of a simple form with great potentiality, rather than a Platonic Ideal that had been debased somewhat in achieving tangible form. And Michael Ruse's (1985) view seems to be that Owen was simply confused about what he was saying. In my view, these commentators are all correct in different respects.

At first, the Archetype was simply a model, a plan on which vertebrates appeared to be built. But in *Limbs,* Owen regarded the Archetype as a kind of Platonic Ideal (see Amundson, this volume). For Plato, the "real" world was the ideal, which we humans can only perceive dimly by their reflections as material objects in our world. It may not matter whether for Owen the Archetype was "real" in an ideal world or the material one. Owen was more interested in showing that concepts such as the Archetype were secondary manifestations of the Prime Mover, one of the law-like phenomena by which the universe operated, under the instigation and superintendence of the Divinity. Such "causes" are most commonly associated with Aristotle. Rupke (1993, 246) thinks that it is reading too much into Owen to expect his philosophy to be consistent, and I would concur. Owen's interest was morphology, not metaphysics; besides, he assembled his biological worldview by synthesizing what he found good and useful from a variety of sources. Why should he have approached philosophy any differently?

OWEN'S QUADROPHENIA

In preparing this essay on the influences that shaped Richard Owen's worldview of morphology and evolution of form, an almost perverse analogy (not homology) seemed to emerge between the structures of Townshend's *Quadrophenia* and Owen's worldview.[1]

1. After the release in 1969 of *Tommy,* the first legitimate opera in the silver age of rock, Peter Townshend, leader of The Who, turned for his next work to his own youth in the lower middle class neighborhoods of London in the late 1950s and early 1960s. Young British "mods" and "rockers" frequently clashed in bars, on streets, and in impromptu or staged gang wars that writ large the personal conflict that Townshend evoked in his masterwork, *Quadrophenia* (1972). "Quadrophenia" was a term that Townshend coined for a condition of mind that pulled his young protagonist in four direc-

Owen's writings show none of the self-doubt, rebellion, insecurity, or indecision of Townshend's protagonist. But just as the latter was faced with opposing forces that he both was drawn to and repulsed by, so Owen felt the pull of influences that he could not accept as his own without rejecting some of their substantial components.

I have outlined above some details of the three cornerstones of Owen's thought, how they fit together, and how he forged them from the works of previous traditions. For his own intellectual "coming of age," which had its first glimmerings in the Hunterian Lectures of 1837 (Sloan 1992; Padian 1995a) and the critical years of 1846–49, when he published *Archetypes and Homologies* and *The Nature of Limbs,* Owen had to define himself in relation to the intellectual spirits of his times, as well as to make peace with some influential ghosts. Some of these relationships are diagrammed in fig. 0.7.

Owen, of course, is at the center of the diagram, which situates him in relation to four different currents of morphological thought important in early Victorian times. Geoffroy St. Hilaire and Georges Cuvier are at opposite ends of one axis; the Oxbridge Paleyites and the German transcendentalists (along with their intellectual descendants in the Edinburgh and London medical communities) polarize another axis. The

---

tions ("Schizophrenic? I'm bleedin' quadrophenic!"): loyalty to his parents, progress at school and at his mundane but decent job, his search for romance and friendship, and the overwhelming desire to fit in with the mod crowd. Self-doubt, guilt, loneliness, betrayal, self-abuse, overindulgence, and lawlessness are only some of his emotions and experiences. In the end he rejects it all in violent frustration, and it is ambiguous whether he does so out of strength of conviction or of despair and collapse. Owen's ambiguity is just as intense, but of a different sort.

Fig. 0.7.

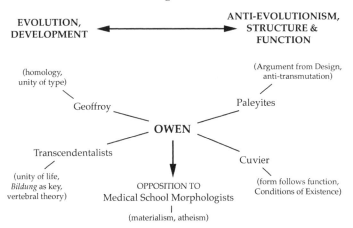

Schematic diagram of the "quadrophenic" relationship
of Owen to four major currents of morphological thought
important to early Victorians, as explained in the text.

respective ends of these axes pair up in another way,
symbolized by the superposed arrow at the top of the
diagram: Geoffroy and the transcendentalists share
an acceptance of evolution in the broad sense and the
conviction of the importance of development in understanding form. Cuvier and the Paleyites reject evolution, and favor the general control of function over form
to notions of underlying unity of plan.[2] At the base of
the diagram, beneath Owen, are notions of materi-
alism and atheism (separate, but linked in Owen's

2. Ospovat (1978) changed the historical model of ideological struggle in these times from the creation-vs.-evolution dichotomy to one
of formalism vs. teleology—in a sense, form vs. function, except
that teleology could have or lack a purpose in Divine design. As he
recognized, no single-axis spectrum is adequate to represent the
complexities of discourse in early Victorian times, and the scheme
pictured here must similarly be a simplification in some respects.

cosmogony) that he consistently despised, whether expressed in Robert Edmond Grant's Lamarckism or, much later, in Darwin's theory of evolution by natural selection.

Geoffroy and Cuvier are well known for their protracted and acrimonious debates on animal form that took place in Paris during 1830 (Appel 1982). Although at the time the perceived outcome was not clear, posterity has generally accepted that Cuvier had the better of the bargain, because he was able to show that there was no basis to Geoffroy's claims of homology between invertebrates and vertebrates (Cuvier in his letters does not even regard Geoffroy as an anatomist, and strictly speaking he is probably best called an embryologist; Le Guyader 2004). However, the association between the two men went back to 1795, when Cuvier first arrived in Paris, and they worked closely and productively for many years. Owen visited Paris in 1831 and was clearly impressed by the grand institution of research and collections that he found at the Museum of Natural History in the Jardin des Plantes; nothing in England was remotely like it. Cuvier was carrying on the tradition of Buffon, the Jussieus, and others, and amplified it with the force of his efforts to turn the science of form into a discipline as lawlike as physics and chemistry had become in the hands of Laplace and Lavoisier. Whereas Owen certainly respected Cuvier's insistence on the interpretation of function from structure, and his ability to predict other aspects of morphology by correlation with preserved parts of fossils, Owen was wary of allowing function too much primacy in shaping morphological form. Perhaps he found an uncomfortable similarity to the Argument from Design that was prevalent among the Paleyites. But there is little question that Owen wanted to establish in England what Cuvier had done in France: not just the elevation of biology to a law-like science—an idea on which Owen would improve in order to meet British sensibilities—but the foundation of an edifice of natural science, a British museum of natural history. Owen, as eventual head of this museum, would be in a comparable position to the nobility with their castles and country estates, and the clergy with their cathedrals.

Geoffroy St.-Hilaire's views on morphology were far closer to Owen's own, though he had to keep this relatively secret, given the abhorrence in some English quarters of transcendental views that were viewed as pantheistic and contrary to teachings about divine intervention in creation (see below). Unlike Geoffroy, however, Owen wanted to retain the teleological, functionalist perspective of Cuvier (Crowe 1985; Richards 1987). Owen knew that much of form is subjugated to function, and that evolutionary change may well occur by organismal response to the exigencies of habitus, a process that Geoffroy would not have endorsed (in part because in Owen's hands this evolution would have been mediated by divine manipulation of archetypal form).

Owen shared with the Paleyites the worship of an omnipotent Creator who, as he explains near the end of *The Nature of Limbs,* foreknew all of his creation. William Paley's *Natural Theology,* the last of many editions of which was published during his lifetime in 1802 (though it is still in print today), encapsuled the precepts of this somewhat middlebrow but broadly appealing religious sentiment, as part of a much broader corpus of works on political and moral philosophy. Briefly, Paley began his argument by imagining that a person crossing a heath was to strike his foot against a stone. He would think nothing of finding the stone, a quite natural part of the environment, in that particular place, nor would he require an extraordinary explanation of how it had come to be there or how it was made. But, Paley continued, if one were to encounter a watch on the same heath, it would be obvious that it was not a natural part of the environment, nor could it have formed spontaneously. It must have been made, because a watch, like any living organism, is a machine, and must be purposefully assembled (Francis 1989; Gillespie 1990). So, he argues, the complexity of living organisms, with their superbly adapted parts, implies a Creator of all life. To deny this is to embrace atheism.

Owen could go as far as divine creation, and he could even accept the wonderful adaptedness of organisms and their parts to their environments and roles in life. But for him, the Argument from Design, particularly as manifested in works such as Sir Charles Bell's (1833) volume in the *Bridgewater Treatises* on the design of the human hand, was what we would today call a "science stopper." That is, once having affirmed that a particular structure is beautifully suited for its role and demonstrates God's magnificence, where does one go with the research plan? Owen preferred, like Cuvier, to look for the natural laws, the Secondary Causes of the Prime Mover, on which a coherent theory of biology could be based. And as his works of 1848 and 1849 show, he believed that he had found it in the Archetype.

But Owen had to be very careful about criticizing natural theology. Paley's latitudinarian views were considered moderate and highly reasonable; many of Owen's patrons among the upper class Tory constituency found Paley logical and comforting; and Owen could not afford to alienate them (Desmond 1982, 1989; Rupke 1995). And so, as Amundson shows in the present volume, he had to endure the slings and arrows of the Puseyites, who wanted him to embrace the Argument from Design and reject the lure of transcendentalism (Richards 1987).

For traditional English theologians, transcendentalist thought was anathema both because it advocated personal reason and intuition about morality and ethics (for example, in Kant's categorical imperative of acting only as one would want all people to act toward each other, as opposed to divine commandment), and because it rejected conventional religion with its emphasis on authoritarianism and ritual. In England, the foremost exponents of transcendentalism were Coleridge, Carlyle, and Wordsworth. For the transcendentalists, God was best known intuitively and through nature, so the unity of humans with nature was central. (Emerson took this idea home with him to America from Paris, where he had an epiphany in the Jardin des Plantes upon seeing the rational taxonomic organization of plants from all corners of the globe and set as his life goal to realize the oneness of humans with nature [Brown 1997].) The ritual and authoritarianism of institutionalized religions repelled them; Coleridge, for example, writing in 1804, could complain in "The world is too much with us" that

> Little we see in nature that is ours;
> . . . for everything, we are out of tune;
> It moves us not. Great God! I'd rather be
> A pagan suckled in a creed outworn . . .

But the transcendentalists were also accused of pantheism, a sort of Victorian code word for nature worship that substituted natural phenomena and laws for divine authority and commandment. For reasons just noted, Owen had to avoid these associations (Desmond 1989; Rupke 1995).

Transcendentalism is generally traced to German morphologists such as Goethe, Schelling, Schiller, Spix, Oken, and Carus, whose work developed out of the Romantic tradition of the late 1700s. Their studies sought an underlying plan in nature, part of which was connected to the transformation of structure. Goethe's studies on plants, for example, established that morphological form is constructed through a process of development, or *Bildung,* during which all parts from leaves to sepals to flowers have a formal correspondence; in a real sense, they are homologous. Goethe, among others, was also a strong proponent of the vertebral theory, which he recalled having recognized while studying a sheep's skull in the Jewish cemetery in Venice (Russell 1916; but Richards [2002] has neatly undercut the probability of Goethe's recollection). The serial correspondence of parts, their "meaning" with respect to an ideal form, is precisely echoed in the opening pages of *The Nature of Limbs* when Owen talks about *Bedeutung.* So Owen drew strongly from the transcendentalists; he is, as Russell (1916, 102) remarked, "the epigonos of transcendental morphology."

In facing each of these four polarizing forces, Owen had to deal with some formidable traditions. The evolution/development strand claimed roots in classical tradition, manifested not only in scholars from Goethe to Oken but in Aristotle himself. In contrast, the antievolution strand was largely rooted in Judaeo-Christian tradition (not, as often wrongly claimed, in the case of Cuvier, who was no biblical literalist), and certainly in the Argument from Design of the Paleyites. In these different strands, knowledge itself was recognized as coming from very different sources (rationality in the first case, revelation in the second), and the explanation of natural phenomena was sought respectively in natural laws and divine plan (Padian 1995b). Owen accepted evolution, certainly in the form of the progression of fossil life through geologic time that had been demonstrated in the rock record. He apparently accepted a kind of evolution of form, based on the Archetype and having to do with alteration of development by the exigencies of habitus (Richards 1987), but his prose was always too vague to suggest anything like testable hypotheses.

WHAT HAS OWEN TO SAY TO US TODAY?

If the answer depends on the validity of Owen's major precepts to modern biology, the answer is surprisingly little, compared to his importance in mid-Victorian times. As Owen scholar Jacob Gruber put it (Gruber and Thackray 1992), Owen simply lived too long. The Darwinians eclipsed his tortuous archetypal philosophy, material causes became more interesting to explain than transcendental ones, and there was no one left to carry Owen's banner or to adapt his ideas to the new age.

What about Owen's cornerstones? The Archetype was forgotten as Darwin (1859) transmutated it wholesale into a Common Ancestor, and tree-thinking began to catch on. The vertebral theory lost its archetypal rationale and was widely discredited, but parts of it, seen through Owen's own lens of serial homology, continue to resonate for many morphologists (e.g., Jollie 1981). The best test of the vertebral theory is Owen's own criteria: position, ontogeny, histology. Neither Owen nor his Darwinian supplanters anticipated the rise of genetics, to say nothing of evolutionary developmental biology.

A major reason why it makes sense to study historical morphologists who sought explanations for the unity of form is that they knew comparative anatomy, morphology, and development so well. In their works, even those that are now centuries old, there is a great deal of useful information; there are problems that they studied and solved that are so much part of textbook lore that we no longer think to remember who did the original work. And even though Owen's rationale for homology and analogy no longer makes sense to us, we continue to need his criteria. After all, it is circular to say that homologous structures are those that are inherited from a common ancestor, and then to posit that we know that the organisms had a common ancestor because the structures are homologous. The question is, what independent evidence determines that these structures are homologous? The answer is in the criteria that Owen used—modified and expanded to accommodate new discoveries in molecular and developmental biology (e.g., Hall 1994).

Darwin and his historical descendants vilified Owen for his hubris, his pettiness, his insincerity, and his malevolence toward his rivals. History is written by the winners, but it was not only the Darwinians that deplored Owen's excesses of personality and his machinations. Biology has rejected his most central ideals and relegated him to near obscurity: ironically, this man who made a career by showing that his intellectual antecedents had not got things quite as right as he could make them was tossed aside because his own grand synthesis failed utterly. But this judgment would overlook the richness and complexity of Owen's views and his accomplishments. In a career that spanned seven decades and resulted in some eight hundred papers, monographs, and books, there are enough contributions to morphology, comparative anatomy, and paleontology to occupy and distinguish several careers. Owen's reputation will endure because of these works, if for no other reason. But beyond that, he was the towering eminence of Victorian biology and paleontology, and without him, it is difficult to imagine the British Museum of Natural History in anything like its present form (Stearn 1981). Owen showed us that it is important, even essential, to have a unified theory of biology, a philosophy that makes the discipline worthy of the name. Owen's principal problem was that Darwin had a very different one that, in the end, was accepted as much superior to his.

ACKNOWLEDGMENTS

I am greatly indebted to James Moore, Adrian Desmond, Ron Amundson, Brian Kraatz, and David Jablonski for many useful comments and suggestions for ideas and references. Of course, they are blameless for any errors, omissions, and misinterpretations.

REFERENCES

Appel, T. 1987. *The Cuvier-Geoffroy debate: French biology in the decades before Darwin*. Oxford: Oxford University Press.

Bell, C. 1833. *The hand: Its mechanism and vital endowments, as evincing design*. London: William Pickering.

Brooke, J. H. 1991. *Science and religion: Some historical perspectives*. Cambridge: Cambridge University Press.

Brown, L. R. 1997. *The Emerson Museum: Practical romanticism and the pursuit of the whole*. Cambridge: Cambridge University Press.

Cadbury, D. 2000. *The dinosaur hunters*. London: Fourth Estate.

Camardi, G. 2001. Richard Owen, morphology, and evolution. *Journal of the History of Biology* 34:481–515.

Crowe, M. J. 1986. *The extraterrestrial life debate 1750–1900: The idea of a plurality of worlds from Kant to Lowell.* Cambridge: Cambridge University Press.

Darwin, C. 1859. *On the origin of species by means of natural selection.* London: John Murray.

Dean, D. 1999. *Gideon Mantell and the discovery of dinosaurs.* Cambridge: Cambridge University Press.

Desmond, A. J. 1979. Designing the dinosaur: Richard Owen's response to Robert Edmond Grant. *Isis* 70:224–34.

———. 1982. *Archetypes and ancestors: Palaeontology in Victorian London, 1850–1875.* Chicago: University of Chicago Press.

———. 1989. *The politics of evolution: Morphology, medicine, and reform in radical London.* Chicago: University of Chicago Press.

Ellegård, A. 1958. *Darwin and the general reader.* Göteborg: Göteborgs Univ. Årsskrift.

Fowles, J. 1969. *The French lieutenant's woman.* Boston: Little, Brown.

Francis, M. 1989. Naturalism and William Paley. *History of European Ideas* 10:203–20.

Gillespie, N. C. 1990. Divine design and the Industrial Revolution: William Paley's abortive reform of natural theology. *Isis* 81:214–29.

Glick, T. F., ed. 1988. *The comparative reception of Darwinism.* Chicago: University of Chicago Press.

Gruber, J., and J. Thackray, eds. 1992. *Richard Owen commemoration.* London: Natural History Museum.

Hall, B. K., ed. 1994. *Homology: The hierarchical basis of comparative biology.* San Diego: Academic Press.

Jollie, M. 1981. Segment theory and the homologizing of cranial bones. *American Naturalist* 118:785–802.

Le Guyader, H. 2004. *Geoffroy Saint-Hilaire: A visionary naturalist.* Trans. by Marjorie Grene. Chicago: University of Chicago Press.

Ospovat, D. 1978. Perfect adaptation and teleological explanation: Approaches to the problem of the history of life in the mid-nineteenth century. *Studies in History of Biology* 2:33–56.

Owen, R. 1843. *Lectures on the comparative anatomy and physiology of the invertebrate animals.* London: Longman, Brown, Green, and Longmans.

———. 1847. Report on the archetype and homologies of the vertebrate skeleton. *Reports of the British Association for the Advancement of Science* 1846: 169–340.

———. 1848. *On the archetype and homologies of the vertebrate skeleton.* London: Richard and John E. Taylor.

———. 1849. *On the nature of limbs.* London: John Van Voorst.

Padian, K. 1995a. A missing Hunterian lecture on vertebrae by Richard Owen, 1837. *Journal of the History of Biology* 28:333–68.

———. 1995b. Form and function: The evolution of a dialectic. In *Form and function in vertebrate paleontology,* ed. J. J. Thomason, 264–77. Cambridge: Cambridge University Press.

———. 1997. The rehabilitation of Sir Richard Owen. *BioScience* 47:446–52.

Paley, W. 1802. *Natural theology.* Repr., Houston: St. Thomas Press, 1972.

Richards, E. 1987. A question of property rights: Richard Owen's evolutionism reassessed. *British Journal of the History of Science* 20:129–71.

Richards, R. J. 2002. *The romantic conception of life: Science and philosophy in the age of Goethe.* Chicago: University of Chicago Press.

Rowe, T. 1987. Definition and diagnosis in the phylogenetic system. *Systematic Zoology* 36:208–11.

Rupke, N. 1985. Richard Owen's Hunterian lectures on comparative anatomy and physiology, 1837–1855. *Medical History* 29:237–58.

———. 1993. Richard Owen's vertebrate archetype. *Isis* 84:231–51.

———. 1995. *Richard Owen: Victorian naturalist.* New Haven: Yale University Press.

Russell, E. S. 1916. *Form and function: A contribution to the history of animal morphology*. Repr., Chicago: University of Chicago Press, 1982.

Secord, J. A. 2000. *Victorian sensation: The extraordinary publication, reception, and secret authorship of vestiges of the natural history of creation*. Chicago: University of Chicago Press.

Sloan, P. R. 1992. Introductory essay and commentary: On the edge of evolution. In *The Hunterian lectures in comparative anatomy, May and June 1837,* by R. Owen, 1–72. Chicago: University of Chicago Press.

Stearn, W. T. 1981. *The natural history museum in south Kensington*. London: Heinemann.

Wells, J. 2000. *Icons of evolution*. Washington DC: Regnery.

Wilson, A. N. 2002. *The Victorians*. London: Hutchinson.

THE MYSTERY OF RICHARD OWEN'S
WINGED BULL-SLAYER

Mary P. Winsor and Jennifer Coggon

The frontispiece of Richard Owen's *On the Nature of Limbs* is a striking but odd cartoon. Who is this winged person, and why is he about to stab a bull? Such a fine pair of wings suggests an angel, but what biblical story does it illustrate? Owen's text provides no hint, and we can assume that his unpublished papers do not contain the answer, or else his scholarly biographer Nicholaas Rupke would have mentioned it.[1] Polly Winsor had long been puzzled by the figure, in a desultory way, but did nothing to solve the mystery until 1996 when, sitting alone in her Toronto office one day, she amused herself by playing Sherlock Holmes. That great fictional detective would have asked, "What clues are sitting right in front of us?" This is a game that forces us to use our imagination, a faculty quite essential to the historian though rarely celebrated. Perhaps the very thing that feels so frustrating to us—that Owen did not identify or explain this image—is itself a clue, she speculated. The frontispiece was surely meant to

---

1. Nicolaas A. Rupke, *Richard Owen: Victorian Naturalist* (New Haven: Yale University Press, 1994).

make his audience comfortable with his topic, and not to perplex them. Owen crafted this text, and its illustrations, for the Royal Institution's Friday Evening Lectures, which aimed to popularize science (although one of his listeners found Owen's lecture too "transcendental").[2] What this implies, Winsor reasoned, is that Owen expected his audience to recognize the winged figure even though we do not. A second clue might be the peculiar unsuitability of the figure. The frontispiece was supposed to show that humans and quadrupeds are morphologically comparable, and to this end Owen carefully labeled homologous bones with the same numbers, and yet this humanoid has bird-like wings. Owen was thereby forced to pretend the wings were boneless, lest this elegant creature be exposed as a four-armed monster. Something about this picture must have seemed to Owen so useful that he forgave the awkward fact of its wingedness. Once again, the implication is that Owen expected his audience to recognize it. What would make a winged-figure-with-bull familiar to an 1849 audience but unfamiliar to us? Winsor guessed that a painting or sculpture seen by Victorian museum-goers might later have been relegated to a basement when tastes changed. Ergo, the place to begin a search for the original of Owen's cartoon would be the catalogues of the British Museum.

This armchair deduction would have gone no further

2. Morris Berman, *Social Change and Scientific Organization: The Royal Institution, 1799–1844* (Ithaca: Cornell University Press, 1978), 125–28; Frank A. J. L. James, "Running the Royal Institution: Faraday as an Administrator," in *"The Common Purposes of Life": Science and Society at the Royal Institution of Great Britain,* ed. Frank A. J. L. James (Burlington, VT: Ashgate Publishing Company, 2002); Gideon Mantell, *The Journal of Gideon Mantell: Surgeon and Geologist, Covering the Years 1818–1852,* ed. E. Cecil Curwen (London: Oxford University Press, 1940), 232.

had not Winsor described it to her student Jennifer Coggon. Traveling to London in the summer of 1998 to do research on another topic (George Newport of frog fertilization fame), Coggon purchased a souvenir for her professor, a glossy booklet introducing the British Museum. It included a portrait of eighteenth-century art lover Charles Townley, surrounded by his collection of Roman marbles, and there, high atop a bookcase in the background, Winsor spied our bull-slayer, tiny but unmistakable.[3] Conjecture confirmed! With the outcome now assured (for all Townley's treasures were acquired by the British Museum in 1804),[4] Winsor assigned Coggon to search through the great illustrated *Description of the Collection of Ancient Marbles in the British Museum*. Success came with volume 10, plate 26, reproduced here as figure 0.8.[5]

Now it is clear that the figure is female, and she is not an angel. She is the goddess Victory, performing the sacrifice of a bull in celebration of the defeat of an enemy.[6] Seeing this accurate and graceful drawing by

3. R. W. G. Anderson, ed., *The British Museum* (London: British Museum Press, 1997), 2.

4. B. F. Cook, *The Townley Marbles* (London: British Museum Publications, 1985), 59.

5. Edward Hawkins, S. Birch, and C. T. Newton, *Description of the Collection of Ancient Marbles in the British Museum, with engravings,* vol. 10 (London: Longman, 1845).

6. Her resemblance to Mithras is striking, confusing, and probably not coincidental. Townley thought his statues, found in Italy, were of a "female Mithras," although that bull-slaying diety sported a flowing cape instead of wings. Hawkins (59–61) recognized the source of the resemblance, and recent scholars agree: Roman sculptors based their carvings of Mithras on a Greek Nike, the same Nike that was the model for their sacrificing Victories. Franz Cumont, *The Mysteries of Mithra,* trans. Thomas J. McCormack (New York: Dover Publications, Inc., [1903] 1956), 210–11; David Ulansey, *The Origins of the Mithraic Mysteries: Cosmology and Salvation in the Ancient World* (New York: Oxford University Press, 1989), 30–31.

Figure 0.8.

© The British Museum

Henry Corbould, published a few years before Owen's lecture, convinced Winsor and Coggon that Owen had based his cartoon on the engraving, rather than making his own drawing of the statue.[7] The only difference between Corbould's and Owen's outlines is that Owen has turned out Victory's right foot, probably to make

7. In 2001 Winsor was able to see the statue itself, now resurrected from storage and on display in a room featuring the Townley collection.

visible her heel, which he numbered to correspond with the bull's hock, and perhaps also to expose her fibula.

At first Coggon's research seemed to confirm Winsor's conjectures, because the entire Townley collection was on proud display in the British Museum in the mid-nineteenth century, but most of it, including the two sacrificing Victories, later went into storage, reemerging only in 1984.[8] Yet Coggon was determined to check into every scrap of evidence available in Toronto's libraries. A newspaper report of Owen's lecture stated that the great anatomist had illustrated his talk with a picture of "a beautiful antique marble."[9] That the reporter used no proper name casts doubt on the notion that this Victory would be instantly recognizable to Owen's audience. Indeed we have found no evidence that this statue made any particular impression on the Victorian public. In fact, the Roman sculpture, which Winsor viewed in 2001, is much less attractive than Corbould's artwork; the marble is stained and stands only 25 inches (0.635 m) in height. We seem to have the kind of case about which Dr. Watson remarked, "Now and again, however, it chanced that even when [Sherlock Holmes] erred the truth was still discovered."[10] For instance, in "Shoscombe Old Place" Holmes is confronted by the fact that Sir Robert Norbert's sister has ceased seeing her friends and keeps to her room. Holmes's guess that Sir Robert has killed the lady leads to the discovery of why her death was concealed, although Holmes also learns that she

8. Cook, *Townley,* 62.

9. "Arts and Sciences: Royal Institution, Feb. 9th," review of Owen's "On the Nature of Limbs" lecture, *Literary Gazette,* February 17, 1849, 113.

10. From the epigraph to Arthur Conan Doyle's "The Yellow Face," (1893), endlessly republished in print and on the Internet.

died of natural causes. In our case it now seems that Owen was not counting on his audience recognizing in his cartoon a sculpture they knew, as Winsor had conjectured; rather, that when he needed an image of a human and animal side by side, Owen made use of an attractive illustration in a recent coffee-table book.

End of story? Not quite, because while Coggon was visiting the University of Toronto Library, bulls kept leaping out at her from the pages of Victorian diaries and popular press. She found herself spellbound. In the late 1840s bulls, both ancient and modern, were a hot topic in London.[11] Coggon read the hair-raising account of a panicked bull that had gored and nearly killed someone, which was not only reported in newspapers but mentioned in the private journal of Owen's wife.[12] That animal had escaped from Smithfield Mar-

11. "Art. VI," review of books and papers on Persian cuneiform, *Quarterly Review* 79, no. 158 (1847): 442, 445, 446; "The Nimroud Marbles," *Athenaeum,* no. 1025 (June 19 1847), 650–51; ibid., no. 1027 (July 2), 706–7; "The Nimroud Antiquities," *Athenaeum,* no. 1098, (November 11 1848), 1128–29; ibid., no. 1099 (Nov. 18), 1152–53; "City Commissioners of Sewers," *Times,* February 14, 1849, 8; "Cattle and Corporation," *Punch,* January 13, 1849, 22; "A Rus in Urbe; or, the Green Hills (Rents) of Smithfield," *Punch,* January 27, 1849, 34; "Sketches of Society: Metropolitan Improvements," *Literary Gazette,* February 3, 1849, 83; [Sara Austen], review of *Nineveh and Its Remains, Times,* February 9, 1849, 5; Elisabeth Fontan, "Adrien de Longpérier et la création du Musée assyrien du Louvre," in *De Khorsabad à Paris: La découverte des Assyriens,* ed. E. Fontan (Paris: Réunion des Musées nationaux, 1994), 230, 232; C. J. Gadd, *The Stones of Assyria: The Surviving Remains of Assyrian Sculpture, their Recovery and their Original Positions* (London: Chatto and Windus, 1936), 126–27; [Henry Hart Milman], "Art. IV," review of *Nineveh and Its Remains, Quarterly Review* 84, no. 167 (1848): 106–53.

12. Richard S. Owen, *The Life of Richard Owen* (London: John Murray, 1894), 2 vols., 1:328–29.

ket, a square in central London where cattle and other livestock were slaughtered.[13] Beginning in 1843, Richard Owen had been active in slaughterhouse reform through his work on the Health of Towns Commission.[14] By late 1849 he was a member of the parliamentary commission that eventually recommended moving the slaughterhouse away from downtown London. Although this commission began a few months after his lecture, the public outcry that led up to it had been building since the late 1820s.[15] The bulls of antiquity were also much in the news, for Assyrian winged bulls had first been uncovered by Paul-Emile Botta in 1843 and exhibited in the Louvre from 1847,[16] and beginning in 1845, the archaeologist-adventurer Austen Henry Layard was uncovering monumental statues in the ruins of what he called Nineveh (really Nimrud). Layard's discoveries included gigantic winged bulls with human heads, and the public eagerly followed sto-

13. Alec Forshaw and Theo Bergström, *Smithfield Past and Present,* 2nd ed. (London: Robert Hale Ltd., 1990), 53–57; Richard Perren, *The Meat Trade in Britain 1840–1914* (London: Routledge and Kegan Paul, 1978), 32–36.

14. Owen, *Life,* 1:216–17; Rupke, *Richard Owen,* xiv.

15. Owen, *Life,* 1:348–49; Perren, *Meat Trade,* 36–37, 40, 225n18.

16. Béatrice André-Salvini, "Introduction aux publications de P. E. Botta et de V. Place," in *De Khorsabad à Paris,* 166–67; "Persian Cuneiform," 438–44; Arnold C. Brackman, *The Luck of Nineveh: Archeology's Great Adventure* (New York: McGraw-Hill, 1978), 204, 206; Fontan, "Adrien de Longpérier," 232, 236; Austen Henry Layard, *Sir A. Henry Layard, G. C. B., D. C. L.; Autobiography and Letters from his Childhood until his Appointment as H.M. Ambassador at Madrid, edited by William N. Bruce, with a Chapter on his Parliamentary Career by Sir Arthur Otway,* 2 vols. (London: J. Murray, 1903), 2:187–88; Gordon Waterfield, *Layard of Nineveh* (London: John Murray, 1963), 114–15, 172, 180, 186; Mogens Trolle Larsen, *The Conquest of Assyria: Excavations in an Antique Land, 1840–1860* (London: Routledge, 1996), 136–37.

ries of their excavation and transportation.[17] Although the formal exhibition of Layard's bulls in the British Museum did not open until 1850, some winged bulls were on display by 1848.[18] By the end of that year, his two-volume book *Nineveh and Its Remains,* with its two striking frontispieces of the removal of a colossal bull from the excavated site, was being reviewed enthusiastically.[19] Coggon was stunned to discover that the Mr. Horne who was recorded in Mrs. Owen's diary as a dinner guest weeks before the "Nature of Limbs" lecture was the same literary man who would later publish a poem in which a Smithfield bull addresses his Assyrian cousin on their mutual maltreatment by humans.[20]

---

17. Ian Jenkins, *Archaeologists and Aesthetes in the Sculpture Galleries of the British Museum 1800–1939* (London: British Museum Press, 1992), 157; Layard, *Autobiography and Letters,* 2:187–88; Shawn Malley, "Austen Henry Layard and the Periodical Press: Middle Eastern Archaeology and the Excavation of Cultural Identity in Mid-Nineteenth Century Britain," *Victorian Review* 22, no. 2 (1996): 153, 154; [Milman], review of *Nineveh and Its Remains;* Julian Reade, *Assyrian Sculpture* (London: British Museum Publications, 1983), 8; Seton Lloyd, *Foundations in the Dust: The Story of Mesopotamian Exploration* (repr., London: Thames and Hudson, 1980), 123–24; Waterfield, *Layard of Nineveh,* 180, 187, 190–93, 197; Brackman, *Luck of Nineveh,* 204–5, 210–12, 220–21, 224; Larsen, *Conquest of Assyria,* 133–37, 141; *Athenaeum* (1847), ibid., 650–51, 707.

18. Gadd, *Stones of Assyria,* 127, 134, 141, 143, 152; Malley, "Layard and the Periodical Press," 155; Larsen, *Conquest of Assyria,* 141; *Athenaeum* (1847), as cited in footnote 11; *Athenaeum* (1848), as cited in footnote 11.

19. Layard, *Nineveh and Its Remains,* 2nd ed., (London: John Murray, 1849), vols. 1 and 2; Layard, *Autobiography and Letters,* 2:191; Malley, "Layard and the Periodical Press," 161–62; [Austen], review of *Nineveh.*

20. Eri Jay Shumaker, *A Concise Bibliography of the Complete Works of Richard Henry (Hengist) Horne (1802–1884)* (Granville,

So now, besides knowing exactly who is the winged figure in the *Nature of Limbs* frontispiece, we have considerable circumstantial evidence suggesting why Owen chose this image. Certainly he was consciously striving to make his lectures attractive to a general audience.[21] He knew, early in 1849, that two kinds of bull, those about to be slain for market and the monumental ones with wings, were in the public's mind.[22] The ambitious Owen, not yet head of the British Museum, understood that his own career advancement depended upon effective communication to nonscientists of the principles of homology that made his subject so impressive to his fellow scientists. The public liked antiquities with wings, they found bloodshed thrilling, and here was a graceful image that combined these elements. Well, what do you think, Dr. Watson, is there any room left for the angels?

ACKNOWLEDGMENTS

We are indebted to Pearce Carefoote of the University of Toronto Library, and to John Dowson of the British Library,

Ohio: Granville Times Press, 1943), 7; Owen, *Life,* 1:218, 333, 348, 361; [R. H. Horne], "The Smithfield Bull to His Cousin of Nineveh," *Household Words* 2, no. 51 (1851): 589–90; Anne Lohrli, *Household Words: A Weekly Journal 1850–1859. Conducted by Charles Dickens* (Toronto: University of Toronto Press, 1973), 309–13.

21. Owen, *Life,* 1: 109–10, 292, 321, 333; [W. J. Broderip and Richard Owen], "Art. III: Generalizations of Comparative Anatomy," review of Richard Owen's works, 1843–1853, *Quarterly Review* 93 (1853): 78–79, 81–82; Richard Owen, *On the Nature of Limbs: A Discourse Delivered on Friday, February 9, at an Evening Meeting of the Royal Institution of Great Britain* (London: J. Van Voorst, 1849), 39.

22. Owen, *Life,* 1:275–76, 352. 2:256–58; Rupke, *Richard Owen,* 135.

for many favors, and to Alan Scollan, of the Department of Greek and Roman Antiquities of the British Museum, who sent us a photograph of the Victory statue. Nathan Sidoli first pointed out to us the Mithras resemblance. Shawn Malley of Bishop's University patiently answered our queries about A. H. Layard. We thank David Williams of London's Natural History Museum for encouraging our pursuit of this topic. When we contacted the Bootmakers of Toronto for advice about Conan Doyle's sleuth, Trevor Raymond kindly put us in touch with Donald Zaldin, who not only supplied the quotation we used from "The Yellow Face," he also showed us that "Shoscombe Old Place" was the best example of Holmes solving a case in spite of his initial guess being wrong. We are very grateful to Ron Amundson for including our story in this volume.

ON

THE NATURE OF LIMBS.

A DISCOURSE

DELIVERED ON FRIDAY, FEBRUARY 9, AT AN EVENING
MEETING OF THE

ROYAL INSTITUTION OF GREAT BRITAIN.

BY

RICHARD OWEN, F.R.S.

LONDON:

JOHN VAN VOORST, PATERNOSTER ROW.

MDCCCXLIX.

οἷον πέπονδεν ὄνυξ πρὸς ὁπλὴν καὶ χεὶρ πρὸς χηλήν.

Aristotle

"Itaque convertenda plane est opera ad inquirendas et notandas
rerum similitudines etanaloga"

Bacon

"Similiter posita omnia in omnibus fere animalibus"

Newton

ON

THE NATURE OF LIMBS.

THE chief difficulty that I have encountered in the endeavour to fulfil the request with which I have been honoured by the Managers of this Institution, is one which the able Secretary will be surprised to hear I owe to him: it has been the attempt to give a plain answer to Mr. Barlow's question, 'By what title shall I announce your Lecture?'

It was not until I had written and erased two or three which first suggested themselves that I became fully conscious how foreign to our English philosophy were those ideas or trains of thought concerned in the discovery of the anatomical truths, one of which I propose to explain on the present occasion in reference to the limbs or locomotive extremities.

A German anatomist, addressing an audience of his countrymen, would feel none of the difficulty which I experienced. His language, rich in the precise expressions of philosophic abstractions, would instantly supply him with the word for the idea he meant to convey; and that word would be 'Bedeutung.' It is the 'Bedeutung' of the limbs which is my present subject; and the literal translation of the word is 'signification.'

B

I had written at first, ' On the Signification of the Limbs of Man and Animals ;' when it occurred to me that a practical audience might deem that there was small need of a learned lecture to prove that matter, and might be disposed to think that any old Pensioner at Greenwich or Chelsea could certify better than a Professor, that ' arms and legs ' signified a good deal.

Our word ' meaning,' as applied to what I shall endeavour to prove, in regard to those useful appendages, would convey as false or feeble an idea of my meaning.

' On the *idea* of the limbs ' might be understood only by those who knew that the word was used in the sense it bears in the Platonic philosophy.

' Homology ' seems now to be accepted as the name of that study or doctrine the subject of which is the relations of the parts of animal bodies understood by the German word ' Bedeutung ;' and in the technical language of anatomical science I should define the present lecture as being : "On the General and Serial Homologies of the Locomotive Extremities." But such a title would have been comprehended only by anatomists, and I knew that your Secretary had in view the information of a more general audience, to whom such technical phraseology would be less intelligible perhaps than the literal translation of the German term.

The ' Bedeutung,' or signification of a part in an animal body, may be explained as the essential nature of such part—as being that essentiality which it retains under every modification of size and form, and for whatever office such modifications may adapt it. I have used therefore the word ' Nature ' in the sense of the German ' Bedeutung,' as signifying that essential character of a part which belongs to it in its relation to a predetermined pattern, answering to the ' idea ' of the Archetypal World in the Platonic cosmogony, which archetype or primal pattern is the basis supporting all the modifications of such part for specific powers and actions in all animals possessing it, and to which archetypal form we come, in the course of our comparison of those modifications, finally to reduce their subject.

The 'limbs' to which the limits of the present Discourse confine its application, are those of the Vertebrate Series of animals; they are the parts called the 'arms' and 'legs' in Man; the 'fore-' and 'hind-legs' of Beasts; the 'wings' and 'legs' of Bats and Birds; the 'pectoral fins' and 'ventral fins' of Fishes. I take for granted that it is generally known, as it is universally admitted by competent anatomists and naturalists, that these limbs or locomotive members, which, according to their speciality of form, have received the above special names, are answerable or 'homologous' parts: that the arm of the Man is the fore-leg of the Beast, the wing of the Bird, and the pectoral fin of the Fish. This special homology has been long discerned and accepted; but the general homology of the parts or their relation to the vertebrate Archetype, in short their 'Bedeutung' or essential nature, is not generally known. Some of the keenest wits and deepest thinkers amongst the anatomists of the German Philosophical School have endeavoured to penetrate the mystery, and have propounded the views which have resulted from such their attempts. But those views have never received the sanction of even partial assent; and, if the conclusion to which I have arrived be the correct one, such assent has not been unreasonably withheld.

It must be owned, however, that the non-acceptance of these generalizations has been due more to indifference and to the non-appreciation of the value of the inquiries, than to a rigorous investigation of their merits. Very few of the exact conclusions as to the general homology of parts of the skeleton of animals have been admitted with thorough comprehension and fruition of the discovery by the actual cultivators of Natural History in this country;

so that I can scarcely appeal to an example in illustration of my meaning with any hope that it will prove such to more than a very small portion of my hearers.

Some however may understand and assent to the proposition that "the basilar part or process of the occipital bone in human anatomy is the 'centrum' or body of a cranial vertebra." Now, by virtue of this truth, mark what the human anatomist comprehends! First, that the '*pars basilaris*' is not a process in the proper signification of the term, but a self-subsistent, independent element of the skull; whereby he is prepared for its primitive appearance as a distinct part in the embryo, and as a persistent distinct 'bone' in all the cold-blooded Vertebrata. He further recognises it to be a member of the same series of bones as the flat bodies of the cervical vertebræ, the thick bodies of the dorsal and lumbar vertebræ, and the broad bodies of the sacral vertebræ; and he understands why it differs from the other 'processes of the occipital bone' by its primitive relation to the embryonic '*chorda dorsalis*.' All this, and much more comfortable knowledge of the '*processus seu pars basilaris ossis occipitalis*,' is implied by the definition of its general homology, *i. e.* of its essential nature, signification, or 'Bedeutung'; and it is precisely the same kind of knowledge which I imply by the word 'nature' in reference to the limbs.

The parts to which I here refer, and to which alone the reasoning will apply that leads to the desired conclusion in the present lecture, are those in the Vertebrated animals, serving chiefly for locomotion, but sometimes adapted to other offices. Many and multiform parts answering these purposes are present in the Invertebrated animals; but their framework is formed out of a distinct system of hard parts from those employed in the Vertebrata. Here it is the internal or endo-skeleton : in the Invertebrata it is the hardened skin, the dermo- or exo-skeleton.

The hard parts of the leg of a Crab or an Insect may be 'analogous' to the bones of the limb of a Quadruped, but they are not 'homologous' with them; and where there is no special homology, there can be no relations of a higher or more general homology between the parts*.

The Vertebrated animals enjoy as extensive and diversified a sphere of active existence as the Invertebrated. They people the seas, and can move swiftly both beneath and upon the surface of water: they can course over the dry land, and traverse the substance of the earth: they can rise above that surface and soar in the lofty regions of aërial space.

The instruments for effecting these different kinds of locomotion—diving and swimming, burrowing and running, climbing and flying—are accordingly very different in their configuration and proportions. The simplest form

Fig. 1.

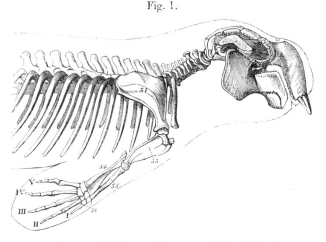

External form and skeleton of the pectoral fin of the Dugong
(*Halicore indicus*).

of the locomotive member is that of the fin. The marine mammal called Dugong here offers us an example of such (fig. 1). It is a strong, stiff, short, broad, flat, and ob-

* The parts termed 'femur,' 'tibia,' 'tarsus,' &c. in Entomology

tusely pointed paddle or oar; without other apparent joint
than that which unites it to the body it has to propel: a
joint permitting that degree of rotation with the oblique
stroke that makes the movement of the oar most effective.

The instrument for burrowing, such as the Mole presents (fig. 2), is not very different in form and character

Fig. 2.

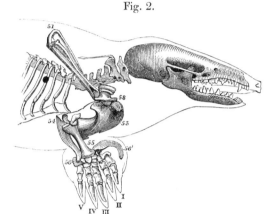

External form and skeleton of the fore-limb of the Mole (*Talpa
Europæa*).

from the fin; but being destined to displace a denser element than water, it is shorter in proportion to its breadth,
and much stronger: it resembles the fin in consisting, seemingly, of but one segment or joint, and being moveable as
a whole only where it is set on to the trunk. The free
border, however, instead of being smooth and thin, is
notched, and armed with a row of hard, tooth-like, horny

were so denominated from a loose appreciation of this analogy, and
convenience obviously suggests their retention in an arbitrary sense.
To attempt to change the application of these names from a supposed
more accurate appreciation of the analogical resemblance, argues either
a mind more subservient to nomenclature than zealous for the advancement of the science of Nature; or an ignorance of the distinct systems
to which the skeletons of the limbs of Articulates and Vertebrates owe
their origin.

points, adapted for scraping and throwing back the soil. With such rapidity does the mole effect this purpose*, that it may literally be said to 'swim through the earth.'

The third form of limb or locomotive member here exhibited (fig. 3), offers a striking contrast to the burrowing

Fig. 3.

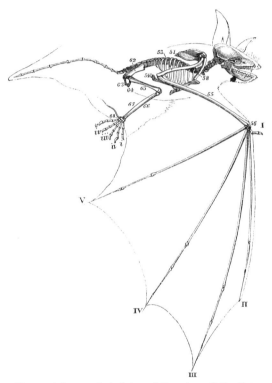

External form and skeleton of the wing of the Bat.

trowel we were last contemplating. It is a thin, vastly expanded sheet of membrane, sustained, like an umbrella, by slender rays, and flapped by means of these to and fro

* Prof. Bell, History of British Quadrupeds, p. 96, 8vo, Van Voorst.

in the air; and with such force and rapidity, as, combined with its extensive surface, to make it react upon the attenuated element more powerfully than gravitation can attract the weight to which the limb is attached, and consequently the body is raised aloft and moved swiftly through the air; in brief, the animal flies, and these instruments of its aërial course are called 'wings.'

When a quadruped has to move swiftly along the surface of the earth by reacting upon the hard ground, its limbs are as remarkable for their length and slenderness as those of the burrower or swimmer are for their shortness and breadth. In the racer the instruments of its rapid course are four long tapering columns, with joints permitting

Fig. 4.

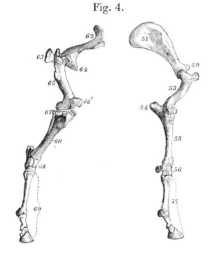

Skeleton of the fore- and hind-leg of a Horse.

them to bend in opposite directions, and of the form represented in the diagram (fig. 4) and familiar to all: each column rests upon a slightly expanded base encased by the hard horny sheath which we call the 'hoof.'

When the limbs are adapted for grasping as well as running, they are divided at their extremities into move-

able appendages or digits, one of which can be opposed
to the others and retain the object of their mutual pressure. Each of the four extremities is so organized in the
ape and monkey, which are thus especially adapted for
climbing and living in trees.

In Man the principle of special adaptation goes further;
and, whilst one pair of limbs is expressly organized for
locomotion and standing in the erect position, the other
pair is left free to execute the manifold behests of his rational and inventive Will, and is exquisitely organized for
delicate touch and prehension, emphatically called 'manipulation.'

Such are some of the more striking amongst the count-
less purposes to which the parts of animals called 'limbs'
are adapted, and such the consequent diversity of their
outward shapes and proportions. We cannot be surprised
at this; it could not be otherwise: the instrument must
be equal to its office. And consider the various devices
that human ingenuity has conceived and human skill and
perseverance have put into practice in order to obtain
corresponding results !

To break his ocean-bounds the islander fabricates his
craft, and glides over the water by means of the oar, the sail,
or the paddle-wheel. To quit the dull earth Man inflates
the balloon, and soars aloft, and, perhaps, endeavours to
steer or guide his course by the action of broad expanded
sheets, like wings. With the arched shield and the spade
or pick he bores the tunnel : and his modes of accelerating
his speed in moving over the surface of the ground are
many and various. But by whatever means or instruments
Man aids, or supersedes, his natural locomotive organs,
such instruments are adapted expressly and immediately
to the end proposed. He does not fetter himself by
the trammels of any common type of locomotive instrument, and increase his pains by having to adjust the parts
and compensate their proportions, so as best to perform

the end required without deviating from the pattern pre-viously laid down for all.   There is no community of plan or structure between the boat and the balloon, between Stephenson's locomotive engine and Brunel's tunnelling machinery: a very remote analogy, if any, can be traced between the instruments devised by man to travel in the air and on the sea, through the earth or along its surface.

Nor should we anticipate, if animated in our researches by the quest of final causes in the belief that they were the sole governing principle of organization, a much greater amount of conformity in the construction of the natural instruments by means of which those different elements are traversed by different animals.   The teleologist would rather expect to find the same direct and purposive adapta-tion of the limb to its office as in the machine.   A deep and pregnant principle in philosophy, therefore, is concerned in the issue of such dissections, and to these, therefore, I now pass, premising that the end in view will be attained without extending the comparison beyond the framework of the limbs, or the leverage of the bones and joints.

The human anatomist is of course familiar with this part of the skeleton in man.   The arm is suspended from a broad bone called 'scapula' (Frontispiece, and Plate I. fig. 6, 51), and the shoulder-joint is strengthened by a slender bone called 'clavicle' (58), which abuts against the top of the 'sternum' (59), forming with its fellow an in-verted arch, called in Comparative Anatomy the 'scapular arch.'   The arm proper is appended to this arch : its first joint or segment is formed by a single long bone, the 'humerus' (53); its second joint, by a pair of shorter and more slender bones, the 'radius (55) and ulna' (54) ; and the hand or third segment is formed by a group of little thick bones, the 'carpals' (56), and by five rays or digits ; one (i) consisting of three segments, the rest (ii—v) of four segments each ; the five bones joining the carpus being called 'metacarpals,' and the others the 'phalanges.'

When we proceed to compare with this the skeleton of the corresponding limb of the horse, ox, or other hoofed animal, the simplification of structure seems not to be in the ratio of the loss of function : almost all that the hoof can be made to do is to rest upon or beat against the ground ; and yet we find in the solidungulate limb (fig. 4) the broad scapula (51), the long humerus (53), the radius (55) and ulna (54), the carpus (56), the metacarpus (57) and the digital phalanges. There is a diminution and simplification of accessory parts, but the essentials are maintained : it is obvious that the same type has governed the formation of the two limbs compared. The most marked distinction is the total absence of the clavicle in the hoofed quadruped : the shoulder-joint did not need to be made the fixed point upon which the fore-limb might rotate in a variety of directions ; on the contrary, the movements of the humerus in the horse, though restricted almost to one plane, are extensive, and the scapula must play backwards and forwards to some extent with the limb ; which movements would have been impeded had it been bound by a brittle bony bar to the sternum.

In like manner, since the hand had no need in the horse or ox to be turned, now prone now supine, in subserviency to manipulation, the accessory joints that allow the radius to rotate on the ulna are abolished, and the ulna is retained only in so far as it is required to strengthen the radius, make the joint of the elbow more secure, and give advantageous attachment to certain muscles. Here, however, we plainly see the same bone, with its olecranon (Frontispiece and fig. 4, 54), although it is anchylosed to the radius (55) and forms what, in Human anatomy, would be called a process of that bone.

The carpal series of small bones answers almost exactly, bone for bone, to that in man* : the hand of the horse,

* Compare fig. 6, *s*, 1, *c*, p, *z*, *m*, *u*, with the carpal bones similarly lettered in the human skeleton, Pl. I. fig. 6, 56.

though to outward view reduced to a single digit, yet shows the rudiments of two others (fig. 6, II and v) when anatomized; these are called 'splint-bones' by the Veterinarians; but we shall presently demonstrate the very fingers in the human hand to which they answer.

If we had little *à-priori* ground to expect so much conformity between the skeleton of the arm of man and that of the fore-limb of the horse, still less have we to anticipate such between these and the bony frame of the widespread wing of the bat. Yet you perceive (fig. 3) that the essential similarity of its composition to that of the human arm (Frontispiece) is greater, the difference depending more on the proportion than on the change or suppression of parts.

Besides the scapula (51) we have the clavicle (58), which reappears in its perfect state; strong, long and curved, rendering the shoulder-joint a firm 'point d'appui' to the vigorous strokes of the wing that centre in the head of the humerus. This bone (53) is proportionally as long as in man, but the bones of the fore-arm are much longer, at least the radius (55); for, rotatory movement of the wrist being here as little needed as in the horse, the ulna (54) is similarly atrophied and reduced to its olecranon and proximal half which is anchylosed to the radius. The small bones of the wrist or carpus (56) again succeed the radius, and here they support the same number of fingers as in man, of which, also, the first or innermost (I) is distinguished from the rest (II—v) by its shortness and difference of direction, and by its inferior number of phalanges. The chief modification of the bones of the fore-limb in the bat consists in the extreme elongation and attenuation of the four ordinary fingers; two or more of which consist, as in man, of the metacarpal bone and of three phalanges, proximal, middle and distal; but the last phalanx tapers to a point and is without a nail.

To skim the air and to burrow in the earth would seem

to require instruments as different in construction as in size and shape; but observe how closely the skeleton of the mole's trowel (fig. 2) conforms in the number and relative position of the parts to that of the bat's wing! The chief change is in this,—whatever is elongated and attenuated in the bat is shortened and thickened in the mole. With regard to one bone, indeed, the statement may be reversed, for the scapula (fig. 2, 51) now appears as a long, straight, prismatic column, and is the centre of a most powerful mass of muscle in the recent limb. The clavicle (58) on the other hand, though perfect, is a short thick cube: the humerus (53) must be classed by its shape amongst the broad bones, the ordinary form of this bone and the scapula being reversed in the mole. The radius (55) and ulna (54) are both completely developed, and enjoy all the accessory rotatory movements as in man; but are relatively much more powerful bones, especially the ulna, which has an enormous olecranon as the fulcrum of powerful muscular forces. Then follow, as usual, the double series of little carpal bones (56), supporting five digits (i—v), which, notwithstanding they are buried up to the claws in a sheath of tough skin, have precisely the same number of bones and joints as in the prehensile hand of man; only that every bone, save the last in each digit, is as broad and thick as it is long. The chief deviation from the human type is by redundancy instead of deficiency, and is exemplified by one or two supernumerary carpal ossicles, the most remarkable of which is sabre-shaped (56′), and strengthens the digging or scraping edge of the broad palm.

If the dissector were little prepared on teleological grounds to meet with the full number of joints or segments in the short and seemingly simple trowel of the mole, he could still less expect to find them hidden beneath the common undivided sheath of the fin of the dugong or whale. Yet the bones of this simple form of limb offer perhaps the most striking and suggestive instance of an adherence to type, necessitated as it would seem, notwithstanding the absence of all those movements and appliances of the limb that explain the presence of the several segments, on the principle of final causes, in the horse and man.

First we have the blade-bone or scapula (fig. 1, 51), now again broad and flat; next the single arm-bone (53), this united by an elbow-joint to a radius (55) and ulna (54); the latter complete with its elbow-prominence or olecranon: the two antibrachial bones are followed by the ossicles of the carpus (56) in their normal double series, and these support the five digits (i–v), with the first distinguished from the rest by its inferior number of phalanges.

In descending to an inferior aquatic species and examining the corresponding (pectoral) fin of a fish (Pl. I. fig. 2), we find indeed one segment (53) of the limb abrogated, and its framework attached to the scapular arch (51, 52) by the two bones answering to radius (55) and ulna (54); these, however, are followed by the constant carpals (56), which support a variable number of digits (57, *a*) in the form of slender pointed rays, like those in the bat, but usually more numerous, often bifurcated, and always divided into a greater number of segments. We clearly recognise, however, the hand, the wrist and the fore-arm in this fish's fin. The number of digital rays in some fishes falls short of the typical 5, and in the *Lepidosiren* (Pl. I. fig. 7) they are reduced to 1, which is partially divided into many segments, and in which all distinction of arm, fore-arm and hand is lost. In the *Murœna* and *Anguis* the scapular arch alone remains, the appended limb being lost; and lastly, in ordinary Serpents and Cyclostomous fishes (Lampreys) all trace of both arch and appendage has vanished.

On glancing back over the great classes of animals represented by the examples that have been adduced, we perceive within what narrow limits of the Vertebrate series the type of the anterior or pectoral member, as shown in man, ceases to be recognizable; and then, not by a change, but by a fading away of the pattern, as the limb rapidly disappears at the extreme of the series. It is by the study of these transitory or rudimental manifestations of the limbs that we gain the deepest and truest insight into their essential nature. But before proceeding to this part of my inquiry, I shall advert to another class of correspondences or evidences of unity of type which give additional impulse to the pursuit of the higher generalization that may reveal the meaning of those evidences.

The bilateral symmetry of the body and the consequent answerableness or parallelism of the parts or organs of one side to those of another, are too obvious facts to need more than a passing reference. They have been universally recognized, were scientifically enunciated by the Father of Natural History*, and in all languages the parts are designated by the same names, and distinguished only by the epithets ' right' and ' left.' Thus we have a right arm and a left arm, a right leg and a left leg, and so with respect to their several segments and the subdivisions of these, down to the right thumb and the left thumb, the right great-toe and the left great-toe, &c. The identity of structure in this transverse direction is complete. The great Newton was so much struck by the contemplation of this law of symmetry, that he breaks out into the following emphatic and beautiful anticipation of the aim and general result of Philosophical Anatomy:—" Tam miram uniformitatem in planetarum systemate, necessario fatendum est intelligentia et consilio fuisse effectam! Idemque dici possit de uniformitate illa quæ est in corporibus animalium. Habent videlicet animalia pleraque omnia, bina latera, dextrum et sinistrum,

---

* " Habet autem homo partes superas et inferas; anteriores et posteriores; dextras et sinistras. Dextræ igitur atque sinistræ partes omnes inter se similes fere et eædem, præterquam quod sinistræ imbecilliores." *Schneider's Aristotelis de Animal. Hist. Lib.* II., tom. ii. p. 30.

forma consimili: et in lateribus illis, a posteriore quidem
corporis sui parte, pedes binos; ab anteriori autem parte,
binos armos, vel pedes, vel alas, humeris affixos: interque
humeros collum in spinam excurrens, cui affixum est ca-
put; in eoque capite binas aures, binos oculos, nasum, os
et linguam; similiter posita omnia, in omnibus fere animalibus."—*Newton, Optices*, 1719, p. 411. Aristotle, also,
saw that, besides the lateral symmetry, the inferior parts
corresponded in a certain proportion to the superior parts.

The most superficial comparison of the limbs of our
own species must impress the observer with the resemblance between the arm and the leg in their general form
and character, notwithstanding the marked contrast of
their powers and offices which characterizes the human
species. Every one may see that the thigh answers to
the arm proper, the leg to the fore-arm, the ankle to the
wrist, the five-toed foot to the five-fingered hand, in which
also the thick thumb may be recognized as answering to
the great-toe, and the little finger to the little toe, and so
of the rest.

When we proceed to the comparison of the bones of the
two limbs, the correspondence becomes more striking. A
broad and flat bone, the haunch-bone or 'ilium' (Frontispiece and Pl. I. fig. 6, 62), with its base towards the spine
and its apex forming the hip-joint, plainly repeats the scapula (51), and the inverted arch is completed by the meeting of two broad and perforated plates below (63 & 64),
which in a general way correspond with the clavicles. The
single bone of the thigh (65) more decidedly repeats the
single bone of the arm (53)*; the two bones of the leg

---

* It will, of course, be obvious that the humerus is not 'the same
bone' as the femur of the same individual in the same sense in which
the humerus of one individual or species is said to be 'the same bone'
as the humerus of another individual or species. In the instance of serial homology above-cited, the femur, though repeating in its segment
the humerus in the more advanced segment, is not its namesake or

(tibia, 56, and fibula, 57) repeat the two bones of the fore-arm (radius, 55, and ulna, 54) ; the small thick bones of the tarsus (68) those of the carpus (56) ; the five metatarsals the five metacarpals : and, if proof were wanting of the serial homology, as I have termed these relations, between the thumb (ɪ) and great toe (*i*), we find it in the skeleton, which shows that both have only two phalanges, whilst the remaining four toes, notwithstanding most of them are shorter than the great-toe, have the same superior number of phalanges, viz. three, as the four fingers have on the hand. The bones that are developed in tendons, which glide over joints, in order to remove them from the centre of motion, are the least regular or subject to type : they are called 'sesamoids,' and the largest and most conspicuous of these is that of the knee-pan or 'patella' (66'), to which there is no corresponding ossification in the homotypal tendon of the arm.

In comparing the bones of the hind-leg (fig. 4, B) with those of the fore-leg (fig. 4, A) of the horse, we find that the modifications by which the latter depart from the type of the skeleton of the human arm are repeated both in kind and degree in the hind-leg. The foot is similarly simplified, attenuated and elongated : it consists chiefly of a single toe (*iii*), with the same number of joints as that of the fore-foot, and touches the ground by a single hoof : rudiments of two other toes (*ii* and *iv*, fig. 19, Pl. I.) are similarly re-presented by the splint-bones : there is a group of tarsal bones (fig. 4, 68) answering to the carpal ones (*ib.* 56). The segment of the leg is formed chiefly by a single bone (66), like the corresponding segment of the fore-limb (53) ; the

'homologue.' I have proposed, therefore, to call the bones so related serially in the same skeleton 'homotypes,' and to restrict the term 'homologue' to the corresponding bones in different individuals or species, which bones bear, or ought to bear, the same names. See 'On the Archetype and Homologies of the Vertebrate Skeleton.' 8vo, V. Voorst, 1848.

fibula (67) being reduced to a rudiment of its upper end, as the ulna (54) is in the antibrachium. The short oblique femur (65) repeats the short oblique humerus (53), only the obliquity is in the opposite sense; but the modification of the end of the bone which produces this change is feeble in comparison to that of the entire femur in different species : compare *e. g.* the femur in man and the megatherium. The long and narrow ilium (62) departs from the human type of that bone just as the long and narrow scapula (51) does from its homologue in man; the chief difference that will be noticed is, that, whereas the coracoid (52) is rudimental and the clavicle *nil* in the fore-limb, their homotypes the ischium (63) and pubis (64) are fully developed, and complete the arch to which the hind-limb is attached. When we remember that the hind-limbs are the chief instruments in propelling the body, we shall see the necessity for the firm and unyielding condition of the part that immediately receives the impetus from the thigh-bone and transmits it to the trunk.

The scapular arch is the seat of more variety in regard to its closed or open state in the class Mammalia than in the lower Vertebrata, which depart less from the type in retaining its complete or closed state. But what chiefly concerns us in the present parallelism between the fore and hind limbs is the fact that the pelvic arch is subject in certain mammals to the same variety by defect as the scapular arch is. Thus, in the skeleton of the bat (fig. 3), in which we saw the scapular arch completed by the extension of the strong clavicles (58) to the sternum, the pelvic arch remains open below, the pubis (64) and ischium (63) being rudimental and forming mere ' processes' of the ilium (62), as the coracoid does of the scapula. Here, therefore, the supporting arches of the fore- and hind-limbs present the reverse conditions to those in the horse in regard to their completeness, and the final cause is obvious. In the bat the fore-limbs are the locomotive instruments, the hind-limbs merely the supporting members from which the animal suspends itself, head downwards.

But if we find that the correspondence between the scapular and pelvic arches is interrupted in the bat to the same extent as in the horse, though contrariwise, the pelvis being open and the scapular arch closed, in relation to the different values of their respective limbs in locomotion; yet the parallelism between the fore- and hind-limbs themselves is equally complete as to their composition. The single femur (65) answers to the humerus (53), and the tibia (66) alone is fully developed in the leg, as the radius (55) is in the fore-arm: whilst the fibula (67) exhibits an equally rudimental and incomplete state to that presented by the ulna (54); only in the leg the lower half of the bone is retained, whilst in the arm it is the upper half: and we may observe a similar contrast in the Camel-tribe. The tarsus (68) in the bat again parallels the carpus (56), and the digits are developed in the same typical number in the foot as in the hand, only that they retain their more normal form and unguiculate character.

There is one item of conformity of structure between the fore- and hind-limbs of certain bats not usually met with in other animals, viz. the development of a sesamoid in the tendon of the *biceps brachii* in front of the elbow-joint, which is the true homotype of the patella in the leg. There is also a second sesamoid above the olecranon (54) of the bat, which has been thought to represent the patella; but I shall presently advert to the true homotype of the ulnar sesamoid in the hind-limb.

The comparison between the fore- and hind-limbs in the mole may be briefly discussed, as they differ in their proportions, not in their composition.

The pelvic arch is reduced in its transverse dimensions to the ordinary size of the hæmal arches in the tail, and surrounds only the pelvic continuation of the aortic arterial trunk; but the ilium resembles the scapula in its length and slenderness: the femur—the homotype of the humerus—is followed by a complete tibia and fibula, answering to the complete radius and ulna in the fore-limb: there is a tarsal segment of small ossicles answering to the carpal series, and five toes with the same number of joints as those of the fore-foot, the first or innermost of each having but two phalanges.

We cannot pursue this parallelism in respect of the limbs that have been adduced as examples of the fin or organ of swimming (fig. 1), because all the whale-tribe are characterized by the absence of the ventral fins or representatives of the hind-legs. We find however on dissection, as in the case of the *Murœna* and slow-worm (*Anguis*), that traces of their sustaining arch still exist. In the whale the ischium or inferior member of the arch is present: in the dugong there is both a rudimental ischium and ilium.

The Reptilian sea-turtle (*Chelone*) is an instance in which both fore- and hind-limbs are modified to serve as fins; but I have selected in illustration of this comparison an extinct form of Reptile, which of all the ancient Sea-lizards (*Enaliosauria*) was most nearly allied to the *Chelonia*. The skeleton of the *Plesiosaurus* (Pl. II.) shows the unity of type closely preserved between the fore- and hind-limbs.

The scapula (51) is a long, narrow, almost vertical bone, both in shape and position like a rib: the arch is completed below by a broad coracoid (52) and a clavicle (58). The ilium (62) in like manner retains much of its proper pleurapophysial or rib-like character, and the pelvic arch is completed below by a broad ischium (63) and pubis (64). The flattened femur (65) closely parallels the flattened humerus (53). In the two short bones of the next segment, the stronger and straighter one in the pectoral fin (55) obviously represents the radius, and its homotype the tibia (66) has the same shape in the ventral fin. The ulna (54) is shorter than the radius, is curved and projects backwards; the fibula (67) repeats the same character: there is a carpal group of ossicles (56) and a tarsal group of ossicles (68); a row of five metacarpals (57), and a row of five metatarsals (69); and the number of the digits, as well as the number of their joints respectively, is almost identical in both fore- and hind-paddles.

The conformity is the more striking here, because it is not broken by the opposite flexures that characterise the corresponding segments of the fore- and hind-limbs in the terrestrial mammalia; and it is the more interesting and significative as being manifested by an animal which lived and died at a period so remote as the existence of that ocean from which the lias of Somersetshire was precipitated.

Vicq-d'Azyr seems to have been the first anatomist whose attention was so much awakened by the perception of these serial correspondences, at least in the human frame, as to have led him to pursue them in detail. He communicated the result of his comparisons to the Parisian Academy, in whose celebrated ' Mémoires' it appears in the year 1774, under the title " Parallèle des Os que composent les Extrémités." The philosophic Condorcet, at that time Perpetual Secretary to the Academy, was so much struck by the grand views of animal structures which this mode of inquiry promised to disclose, that in his Report on the Memoir he characterises it as ' a new kind of Comparative Anatomy.' It seldom happens, however, in such early excursions into a foreign territory, that success is complete; and Vicq-d'Azyr, though right in respect of the scapula, humerus, and bones of the hand, was led astray by the different proportions of the ulna and fibula in the human skeleton in his comparison of those bones.

Preconceptions from modifications of *form* can only be corrected by an extended survey of such variable morphological characters. Vicq-d'Azyr conceived that the ante-

rior extremity was not paralleled or repeated by the posterior one of its own side, but by that of the opposite side;
the right arm by the left leg, and *vice versâ*; and Cuvier
adopted this notion of a dextro-sinistral conformity*.

Vicq-d'Azyr also supposed that the ulna in the fore-
arm was the correspondent of the tibia in the leg, with the
patella as the detached olecranon, and that the radius
answered to the fibula; a mistake which he could scarcely
have fallen into had he extended his comparisons to the
lower Mammalia, or known the instances which I shall
presently adduce from the Marsupial order.

So tardily however have the exact generalizations of
Philosophic Anatomy been appreciated in France, that, in
one of the latest and most elaborate ' Manuals of Descriptive Anatomy,'—a work of considerable repute for its
minute accuracy of detail,—it is laid down, that " the upper end of the tibia is represented by the upper half of the
ulna, and the lower half of the tibia by the lower half of
the radius; whilst the fibula is represented by the upper
half of the radius and the lower half of the ulna†."

Nature, however, when rightly interrogated and propi-
tiated by due observant service, extricates us from these
complex involutions and alternations of serial homology,
and makes the simple truth plain. The extensive knowledge of Comparative Anatomy possessed by my revered
preceptor in Anatomy, Dr. Barclay, enabled him truly
to interpret the parallelism of the bones of the fore-arm
and of the leg-proper. He showed how the ulna and its
homotype the fibula exhibited the same ' variety and un-
steadiness of character, sometimes large, sometimes small,

* Leçons d'Anat. Comp. t. i. 1836, p. 342.
† " L'extrémité supérieure du tibia est représentée par la moitié supérieure du cubitus, et la moitié inférieure du tibia par la moitié inférieure du radius; tandis que le péroné est représenté par la moitié
supérieure du radius et par la moitié inférieure du cubitus."—*Cruvel-
hier, Anatomie Déscriptive*, t. i. p. 315.

and sometimes merely a process' of the more constant bone of their respective segments*.

Some anatomists may still be biased by the cause of Vicq-d'Azyr's mistake, viz. the great development of the olecranon or upper end of the ulna: but it needs only instances in which the fibula manifests a similar development to satisfy the most sceptical as to the soundness of the grounds on which the Homological conclusions illustrated in Plate I. figs. 15 and 16 are based. The marsupial and monotrematous animals are fertile in this parallel instance of excessive development. The head of the fibula in the ornithorhynchus extends far beyond the knee-joint, and is expanded like the olecranon in the mole's ulna. Many of the marsupial quadrupeds have a rotatory motion of the hind-foot analogous to the pronation and supination of the hand: in the opossums and phalangers the great-toe is an opposable thumb, whilst its homotype in the hand remains parallel with the other fingers, and the location of the foot and hand is thus the reverse of that in the human subject: whence naturalists have styled these marsupial animals ' *Pedimana* '—foot-handed. One might expect that the modifications of the

---

* M. Flourens had probably never seen Dr. Barclay's ' Explanations of Mitchel's Plates of the Bones,' 4to, 1824, when he wrote, "Il a été plus difficile de rapporter individuellement chaque os d'un membre à chaque os de l'autre. Chose étrange, on ne sait pas encore s'il faut comparer ensemble l'*humerus* et le *femur* du même côté ou l'*humerus* d'un côté et le *femur* de l'autre ; on ne sait pas quel est celui des deux os de l'*avant-bras*, le *radius* ou le *cubitus*, qu'il faut comparer à tel ou tel des deux os de la jambe, le *tibia* ou le *péroné*." He supports his reproduction of Barclay's proposition regarding the serial homology of the bones of the fore-arm and leg by similar remarks drawn from Comparative Anatomy. " Déjà," writes M. Flourens, " dans les Chauve-souris, dans les *Galéopithèques*, le *cubitus* n'est plus qu'un filet très grêle ; ce même *cubitus* ne se montre plus qu'en vestige dans les *Ruminans*, dans les *Solipèdes* ; le *péroné*, déjà très grêle dans les Chauve-souris, déjà simple rudiment styloïde dans le *Cheval*, manque à-peu-près tout-à-fait dans plusieurs *Ruminans*."—*Annales des Sciences Naturelles*, 1838, pp. 35, 37.

bones of the leg, in subserviency to those transferred powers of prehension, would resemble the characters that disguise the serial homology of the bones of the fore-arm in man ; and accordingly we find that, whilst the co-adapted joints between the tibia and fibula in the *Pedimana* are such as to permit their reciprocal rotation, the proximal end of the fibula is developed to afford the requisite advantage to the muscles acting upon the foot.

In illustration of this instructive parallelism I have selected the leg-bones of the wombat (*Phascolomys*, Pl. I. figs. 15 and 16), because in them not only does the fibula (67) show the broad, flat and high process from its proximal end, but also a superadded sesamoid (67') which parallels the peculiarity observable in the olecranon of the bat (fig. 3, 54), viz. the separate superincumbent ossicle, which at first sight might have been deemed confirmatory of V.-d'Azyr's idea of the patella being a detached olecranon.

In instituting the same kind of comparison between the little bones of the carpus and tarsus, I have been forcibly struck with the perspicacity with which V.-d'Azyr detected their true homologies in the human skeleton, notwithstanding their difference in form and number. He compares the 'scaphoides' (*sc*) of the wrist (Pl. I. fig. 6, 56) with the 'scaphoides' or 'naviculare' (*s*) of the ankle (*ib.* 68), the 'lunare' (*l*) with the 'astragalus' (*a*), the 'cuneiforme' (*cn*) and 'pisiforme' (*p*), together, with the 'calcaneum' (*cl, cl*) ; and, in the second row, the 'trapezium' (*t*) with the 'entocuneiforme' (*ci*), the 'trapezoïdes' (*z*) with the 'mesocuneiforme' (*cm*), the 'os magnum' (*m*) with the 'ectocuneiforme' (*ce*), and the 'unciforme' (*u*) with the 'cuboïdes' (*b*).

To some anthropotomists, viewing the different position of the scaphoid in the wrist (*sc*) and ankle (*s*) of the human skeleton (Pl. I. fig. 6), these comparisons may still seem forced and fanciful : and since the calcaneum (*cl, cl*) is actually developed from a single ossific centre, they may regard its division as purely arbitrary, in order to form the homotypes of the two outer bones of the proximal row of carpal bones. And these objections can only be explained and removed by reference to an enlarged survey of all the modifications of the bones in question.

As M. Flourens, in adopting the determinations of Vicq-d'Azyr, has not met these obvious and seemingly natural objections, by reference to Comparative Anatomy, I shall adduce some of those instances which illustrate the true character of the carpal and tarsal bones, especially in regard to the classification of the bones of the skeleton into 'simple' and 'compound*.'

Without looking further than the human skeleton, it will be obvious that the distal row of the tarsal bones is undisturbed (Pl. I. fig. 6, *ci*, *cm*, *ce*, *b*), and in the same relation to the metatarsals as the distal row of the carpals (*t*, *z*, *m*, *u*) is to the metacarpals: the cuboid (*b*) for example supports the two outer toes, as the unciforme (*u*) supports the two outer fingers. But the same order does not prevail in the arrangement of the other bones of the tarsus: the scaphoid (*s*) is so displaced as to represent part of an intermediate row. M. Flourens explains this modification by reference to the necessity of the greater length required by the inner digit of the foot for the functions of that member, which elongation he affirms to be exclusively caused by this displacement of the scaphoid†. A glance, however, at the relations of the advanced scaphoid in the human skeleton (*s* in fig. 6, Pl. I.) will show that it ought equally to affect the length of the toes attached to the

---

* 'On the Archetype and Homologies of the Skeleton,' 8vo, pp. 103–105.

† "Or, supposez le *semilunaire* grandi à la main, comme l'*astragale* l'est au pied, il repoussera nécessairement le *scaphoïde*, il le portera en avant; et, ce qui le prouve, c'est l'allongement du pouce du pied, comparé au pouce de la main, allongement qui n'a, en effet, d'autre cause que le déplacement du *scaphoïde*, son transport en avant, et sa position sur la même ligne que les autres os du pouce."—*Annales des Sciences Naturelles*, t. x. p. 39, 1838.

mesocuneiform (cm) and ectocuneiform (ce) bones; and
that the elongation of the great-toe is due rather to the
disproportionate length of the entocuneiform (ci) or homotype of the little trapezium (t) of the wrist. The true explanation, therefore, of the difference in question must be
sought for in other and wider considerations.

I have elsewhere remarked* that, " In mammalian quadrupeds generally the fore-limb takes the greater share in
the support, the hind-limb in the propulsion of the body.
The *manus* is accordingly commonly shorter and broader
than the *pes*; this may be seen in the terminal segment of
even the monodactyle hand and foot of the horse. Consequently the transverse direction prevails in the arrangement of the carpal bones and the longitudinal in that of
the tarsal bones. The difference is least in the carpus and
tarsus of the long and slender fore- and hind-hands of the
Quadrumana. If the carpus of the chimpanzee, for example, be compared with that of man, the first difference
which presents itself is the comparatively small proportion
of the scaphoid which articulates with the radius, as compared with that in man, in whom the distal articulation of
the radius (Pl. I. fig. 6, 55) is equally divided between the
scaphoides (sc) and lunare (l) which are on the same parallel transverse series. In the orang (Pl. I. fig. 13), the
divided scaphoid (s, s') extends, almost as much from the
lunare as from the radius, along the radial side of the
carpus, to reach the trapezium (t) and trapezoides (z); it
is in great part interposed between the lunare (l) of the
proximal row and the trapezium and trapezoid of the distal row of the carpal bones. The similarity of its connections, therefore, in the carpus with those of the scaphoid
in the tarsus (Pl. I. fig. 14, s) is so close that the serial
homology of the two bones is unmistakeable. The astragalus (*ib. a*), then, in the foot, repeats the os lunare (l) in
the hand, but usurps the whole of the articular surface of

* 'On the Archetype and Homologies of the Skeleton,' p. 167.

the tibia, and presents a larger proportional size, especially in man, whose erect position required such exaggerated development of the astragalus, or homotype of the lunare. The prominent part of the calcaneum (Pl. I. figs. 6 and 14, cl') obviously repeats the prominent pisiforme (figs. 6 and 13, p), and the body of the calcaneum (figs. 6 and 14, cl) articulates with the fibula, as the cuneiforme (figs. 6 and 13, cu) articulates with the ulna. The strain upon the homotype of the pisiforme (cl') to produce the required effect in raising the back-part of the foot with its superincumbent weight upon the resisting ends of the toes, required its firm coalescence with the homotype of the cuneiforme." The calcaneum, therefore, is essentially a 'compound bone'; that is, it answers to two bones, and includes them ideally, though they be connate*. We might infer the same in respect to the unciform (v) in the hand and the cuboid (b) in the foot, seeing that they each support two digits, whilst the other ossicles of their series respectively support a single digit.

The test of the truth of this idea will be the result of a comparative survey of the carpal and tarsal bones in the Vertebrate series : and, accordingly, in descending to the cold-blooded animals that are more obedient to the archetypal law, and where more of the primitive ossific centres continue distinct, we find in the Chelonian reptiles for example (Pl. I. fig. 12) the unciform bone represented by two bones (u, u'), and each of the five metacarpals is supported by its own carpal ossicle. This structure naturally suggests that the normal or typical number of carpal bones is ten, viz. five in each row, corresponding with the typical number of the digits. But we should search in vain the

---

* This term is used in the definite sense explained in my work on the 'Archetype of the Vertebrate Skeleton' (8vo, V. Voorst, p. 49), as signifying those essentially different parts which are not physically distinct at any stage of development; and in contradistinction to the term 'confluent,' which applies to those united parts which were originally distinct.

human carpus to find any characters that would justify our choice of the compound or essentially double bone in the proximal series, like those that indicate the compound bone in the distal series. Comparative Anatomy, however, affords the key to this problem and directs our choice. The same instance (Pl. I. fig. 12) which gives the typical bipartite condition of the unciforme (u, u') also presents it in the 'scaphoïdes' (s, s') : and we become assured that the second ossicle (s') is no mere accidental and exceptional supernumerary from the frequent repetition of the divided 'scaphoid' in higher classes : it reappears, for example, in the Quadrumana, and the typical character of the bone is even retained in the species which approaches so near to man as the orang-utan (Pl. I. fig. 13, s, s'). The only difference from the tortoise being, that whereas in that reptile the two scaphoids in the wrist articulate with the three carpal bones of the second row, answering to the three cuneiform bones of the ankle, and thus repeat the connections of the undivided tarsal 'scaphoid'—in the orang the divided carpal 'scaphoid' supports only the trapezium and trapezoïdes, answering to the inner and middle cuneiform bones, the os magnum (m) being so elongated as directly to articulate with the lunare (l). In both the tortoise and orang, however, we perceive that the arrangement of the five bones of the proximal row of the carpus is irregular, and parallels that of the corresponding bones of the tarsus ; the two bones represented by the scaphoid intervening between the distal row and those bones that remain in the proximal row.

With what new interest must the human anatomist view the little ossicles of the carpus and tarsus when their homologies have been thus determined ! It must be evident to him that their true nature could never have been understood by the study of them in the human skeleton alone, however minutely scrutinized there. But by the light reflected from Comparative Anatomy he is now enabled both to discern their homotypal relations and their natural classification. In the carpus, the lunare, cuneiforme, pisiforme, trapezium, trapezoïdes, and os magnum rank with the 'simple' bones; and the scaphoïdes and unciforme with the 'compound' bones. And so likewise in the tarsus, the calcaneum and cuboïdes are the compound bones, the others are simple bones. Connation or blending together of two essentially distinct bones is by no means however confined to the scaphoid and unciforme in the wrist, or to the calcaneum and cuboid in the ankle.

In many ferine, rodent, and marsupial quadrupeds the scaphoid unites with the lunare ; examples of such scapholunar bone are shown in Pl. I. figs. 5 and 15, *sc, l.* In the sloth and megatherium the trapezium blends with the scaphoid. In the tarsus of the ruminants the cuboid is anchylosed to the scaphoid : an example of this scaphocuboid bone is given in the hind-foot of the ox (*s, b,* fig. 18, Pl. I.), and other instances might be multiplied. Those that have been adduced may serve, however, to remind the hesitating or sceptical anatomist, that the same reasons which prevailed with Cuvier to recognise the compound character of the scapho-lunar bone of the lion, the scapho-trapezial bone of the sloth, and the scapho-cuboid bone of the ox, have led me to the same conclusion with regard to the unciforme in the carpus, the calcaneum in the tarsus, and the bone called scaphoid in both those segments of the limbs of man.

Another important and instructive result of the foregoing comparisons is the constancy of the relations of the distal series of carpal and tarsal bones, whether simple or compound, with the five digits with which they essentially correspond in number : for by this constancy of connexion we are able to determine the precise digits that are lost and retained when their number falls below the typical 5 ; to point out, for example, the finger in the hand of man that answers to the fore-foot of the horse, and the toe that corresponds to its hind-foot; nay, the very nail in the hand or foot which becomes by excess of development the great hoof of the horse. Were anything wanting to impress the thinking mind with the conviction of the unity of type which pervades animal structures, it might be such a fact as this.

The small styliform ossicle which is attached to the trapezium in the wrist of the spider-monkey (*Ateles*) or the hyæna, is plainly shown by that connexion, besides its relations to the other digits, to be a remnant of the thumb—the first digit of the hand that disappears in the process of reduction. The similar ossicle that is attached to the diminished unciforme of the marsupial bandicoot, in like manner is shown by that connexion to be the rudiment of the little finger; the three remaining digits also retaining respectively their normal connexions with the trapezoïdes, the magnum, and the unciforme.

In the tridactyle rhinoceros a mere rudiment of the trapezium lingers in the wrist, but the homologue of the thumb is lost; and the part of the compound unciforme supporting the little finger has disappeared with every trace of that digit.

In the ox we find a rudiment of the fifth digit (fig. 5, v) attached to the outer part of the carpal bone (*u*) articulating with the outer half of the cannon-bone (iv); and thus we recognize the os unciforme supporting, besides that rudiment (v), the metacarpal of the fourth digit (iv), which has coalesced with that of the third digit (iii): the inner half of the cannon-bone, representing the third or middle metacarpal, articulates, in fact, with a distinct carpal ossicle (*m*), which, by its relations to the unciforme (*u*) externally and to the lunare (*l*) and scaphoïdes (*s*) above, is plainly the homologue of the os magnum (*m*). A rudiment of the second metacarpal (ii) is retained on the inner side of the base of the cannon-bone, and a trapezoid is distinct in the carpus of the camels and the diminutive chevrotains (*Tragulus*) ; but it has coalesced with the magnum in the ox, and both trapezium and thumb have disappeared in all Ruminants. Thus we learn that the fully developed digits that support the cloven hoof of the Ruminants answer in the fore-foot to the middle and ring fingers of our hand, and that the cannon-bone consists of the coalesced metacarpals of those two fingers.

Fig. 5.   Fig. 6.

Bones of the fore-foot of the   Bones of the fore-foot
Bos Urus.   of the Horse.

When we examine the bones of the fore-foot of the horse (fig. 6), we find the cannon-bone (III) articulated to a single bone of the carpus (*m*), which, from its relations to the scaphoid (*s*) and lunare (*l*), is the homologue of the os magnum in the ox; but it is relatively of larger size, having increased to the dimensions required for its application to the whole of the proximal surface of the great metacarpal (III). This is not, however, the equivalent of the two united bones (fig. 5, III and IV) so placed in the ruminant, for the unciforme (fig. 6, u) exists external to it, but reduced in size in conformity with the reduction of the rudimental metacarpal (IV) which it now supports. The carpus of the horse, in fact, resembles that of the rhinoceros more than that of the ox ; and the readiest key to the nature and homologies of the bones of the horse's foot is afforded by comparison with that of the ponderous three-toed Pachyderm. The rudiment of the trapezium has disappeared in the horse, but the three other bones remain, and the middle one is the largest, as in the rhinoceros. The extinct Palæotherium offers a connecting link in the transition to the apparently monodactyle foot. The trapezoïdes (z) and unciforme (u) are however still more reduced in size, in the horse, corresponding with the reduction of the toes, which they support, to mere styliform remnants of their metacarpal bones ; and the os magnum and its metacarpal bone are proportionally increased in bulk.

Thus not only are the first and fifth digits wanting in the horse, but the second and fourth are rudimental, and only the third is retained complete, and it constitutes almost the entire foot. The cannon-bone of the horse is not therefore composed, like that of the ox, of two metacarpals confluent, but is essentially as well as actually a single bone ; and accordingly it supports a single toe of three phalanges, which, in the special language of the Hippotomist, are the 'great pastern-bone' (1), the 'small pastern-bone' (2), and the 'coffin-bone' (3) ; the little sesamoid behind the last joint being called the 'navicular' or 'nut' bone *.

* The common term 'cloven-hoof,' applied to the foot of the Ruminants, indicates the idea that it was the solidungulous foot split up ; and the converse in respect of the horse's foot has been taught by high and justly esteemed authorities. Sir Charles Bell, in his 'Bridgewater Treatise,' 8vo, 1833, states, p. 82, " In the horse, the cannonbone may be shown to consist of two metacarpal bones." But he does not give the demonstration. He refers (p. 85) to a comparison which led him to view the phalanges as still more compounded. " In looking

A perfect and beautiful parallelism reigns in the order in which the toes successively disappear in the hind-foot with that in the fore-foot. The 'hallux' or inner toe, answering to the great-toe in man, is the first to fade away : e. g. in the dog (Pl. I. fig. 5), where its rudiment may be found (i 1, 69) with a slender entocuneiform (c i) : in the hippopotamus both the digit and its tarsal bone are wanting. The outer or fifth toe, answering to the little toe in man, is reduced to a rudiment in the alactaga, and totally disappears in the tapir and rhinoceros, where fig. 17, Pl. I. shows the middle toe in its normal relation with the ecto-cuneiform, the second toe with the mesocuneiform—now the internal bone, and the fourth toe with the reduced cuboides. Fig. 18 shows that the toe which next disappears, e. g. in the hind-foot of the Ruminant, is the second, as in the fore-foot; the third and fourth being retained to support the 'cloven-hoof'; the one articulating with the ectocuneiform (c e), the other with the cuboid (b), here confluent with the scaphoid. The rudiments of the second (ii) and fifth (v) digits appear externally as the 'spurious hoofs,' which dangle or project behind the normal one. These, however, are not without their use: when the elk or bison treads in swampy ground, the hoofs expand, the false hoofs are pushed out, and the resisting surface is increased as the

to this sketch and comparing it with that of the hand on page 79, we see that in the horse's leg the five bones of the first digital phalanx are consolidated into the large pastern-bone ; those of the second phalanx into the lesser pastern or coronet ; and those of the last phalanx into the coffin-bone." The learned Professor of Comparative Anatomy in University College, London, describing the bones of the horse's foot, also says, : " We observe that the phalanges, three in number in each toe, of the anterior and posterior extremities, are composed each of two bones anchylosed together, so that only one toe appears to touch the ground, which is covered with a large undivided hoof, from which they are called 'Solidungula'." And again : " The anchylosis seen in the cannon-bone of the Ruminantia has here proceeded downwards through the whole extent of the feet."—Dr. Grant's Lectures, 'Lancet,' No.550, March 1834, p. 907.

foot sinks; but when it is lifted up the small hoofs collapse to the sides of the large ones, which contract, and by this diminution of the size of the foot the act of withdrawal is facilitated*.

In the Ruminants confined to arid deserts we should hardly expect to meet with the mechanism which seems expressly adapted to the marsh and the swamp; and in fact every trace of the second and fourth digits has disappeared from the feet of the camel and dromedary. The comparison of the bones of the extremities is replete with these beautiful evidences of design; but our present purpose is to gather the indications of that which has been sometimes, but wrongly, regarded as the antithetical principle, viz. the unity of plan which lies at the bottom of all the adaptive modifications.

In the hind-foot of the horse you will perceive that the homologue of our small external cuneiform bone (*ce*, fig. 6, Pl. I.) has attained an excessive size in fig. 19, *ce*, and that the toe (*iii*) which that bone supports throughout the Mammalian series is correspondingly developed; whilst the mesocuneiform (*m*) and the cuboid (*b*) are in an equal degree reduced, and, like their carpal homotypes the trapezoid and unciform, they support only rudiments of the second and fourth metatarsals in the form of the accessory splint-bones (*ii* and *iv*).

To sum up, then, the modifications of the digits: they never exceed five in number on each foot in any existing vertebrate animal above the rank of Fishes; and in the class *Mammalia*, the *Cetacea* excepted, the number of phalanges is limited to two in the first digit and to three in each of the other digits, in both fore- and hind-feet. In the cuts and in each figure of Pl. I. they are numbered from the innermost, I, II, III, IV, V, in the fore-foot, and *i, ii, iii, iv, v*, in the hind-foot: these are their symbols, and are

* Sir Charles Bell has well shown the advantage of the cloven hoof over the undivided hoof in this respect, *op. cit.* p. 89.

applied arbitrarily to their objects; thus the digit (*ii*) in the rhinoceros is the innermost or first of its series, and the digit (*iv*) is the third; but the notation signifies in this and all the figures that the digits *ii, iii,* or *iv,* are those answering to the second, third and fourth in the fully-developed foot.

The first or innermost digit, as a general rule, is the first to disappear; in the hind-foot of the orang (fig. 14) commonly, and in that of the wombat (fig. 16) constantly, its short metatarsal supports but one phalanx; in the dog (fig. 4), the inner digit is usually wanting in the hind-foot, and is always very diminutive in the fore-foot. The first digit of the hand is reduced to a short metacarpal in the spider-monkeys (*Ateles*).

The outer digit v and *v* is the next to disappear. In the tapir it is wanting in the hind-foot; and in the rhinoceros (fig. 17) in both hind- and fore-feet.

In the bisulcate quadrupeds the development of the second digit (*ii*) is arrested in addition to the outermost one (v), and the functions of support and progression are committed to the equally and symmetrically developed digits *iii* and *iv*: rudiments of the second and fifth digits are retained in most Ruminantia (as at *ii* and *v,* fig. 18); but in the camel-tribe they have entirely disappeared, together with the first digit, ı and *i.*

In the horse (fig. 19) the fourth digit is the additional subject of arrested development, and the median one in both fore- and hind-feet, ııı and *iii,* is the last and sole digit which retains its full and functional perfection, thus manifesting its character as the most constant and essential of the terminal ramifications of the primitive ray which we saw in the lepidosiren.

Whilst the number of toes is thus seen to fall short, progressively, of five, the typical character of that number is still indicated by the power of determining the particular toe or toes of the five in man, which are retained in the tetradactyle, tridactyle, didactyle and monodactyle feet respectively of the lower mammals. But although the number 'five' thus governs the development of digits, properly so called, in all existing air-breathing Vertebrata, the tendency to multiplication of terminal rays in the diverging appendages modified for locomotion may be seen to manifest itself in the sexual 'spurs' of the Gallinaceous birds and Monotremes; in the hereditary supernumerary toes in certain varieties of the common fowl, and even in some individuals of the human race. But the single spur of the tetradactyle cock is not more a homologue of a normal digit in a pentadactyle reptile or mammal, than is the spur of the *Platypus*, or either of the spurs in the *Pavo bicalcaratus**.

So long as the digits are developed as simple rays they are not subordinated to the typical number, but usually much exceed it, as we find in most fishes. In the skate (*Raia*), indeed, the pectoral members far surpass in bulk and seeming complexity their homologues in man: but their development is of a lower kind: it consists of a vegetative repetition,—division, bifurcation and segmentation— of mere rays, of a multiplication of essentially similar parts, without power of reciprocal action and reaction on one another; all being bound up in one common fold of integument for one simple kind of flapping motion—the only one required for an animal so low in the scale, but perfectly provided for by the form of fin in question. At first sight the pectoral fin of the skate with its hundred digits seems a more complex deviation from the primordial single ray, as shown in the lepidosiren (Pl. I. fig. 7), than the pentadactyle upper extremity (fig. 6, 53—57) of man; but this is far from being the case: true complexity is not shown in the number, but in the variety and coordination of parts.

The high characteristics of the human arm and hand are manifested by the subordination of each part to a harmonious combination of function with another, by the de-

---

* These horny appendages of the metatarsus appear rather to be homotypes of the strong quills attached to the metacarpus in birds.

parture of every element of the appendage from the form
of the simple ray, and each by a special modification of its
own; so that every single bone is distinguishable from another: each digit has its own peculiar character and name,
and the 'thumb,' which is the least constant and important
of the five divisions of the appendage in the rest of the class,
becomes in man the most important element of the terminal
segment, and that which makes it a 'hand' properly so
called.

In the pelvic, as in the scapular extremity, the same
digit (i), which is the first to be rejected in the mammalian
series, becomes, as it were, 'the chief stone of the corner,'
and is termed 'par excellence,' the 'great-toe:' and this is
more peculiarly characteristic of the genus *Homo* than even
its homotype the thumb; for the monkey has a kind of
pollex on the hand, but no brute mammal presents that
development of the *hallux*, on which the erect posture
and gait of man mainly depend.

We perceive, however, that although the first toe (Pl. I.
fig. 6, i) is the longest as well as the largest, it retains its
characteristic inferior number of phalanges; its bulk depending, like the larger toe in the didactyle ostrich (fig. 11,
iii), on the superior size instead of an increased number of
bones; whilst the fifth or little toe (v) still retains with
diminished proportions its full complement of phalanges.
The teleologist will discern that the requisite strength of the
toe, which is the chief fulcrum when the whole body is raised
by the power acting on the heel, as in stepping forward,
has been regarded in the diminished number of its joints;
but the same final cause would not appear to have governed
the different number of joints of the equally-sized first and
fifth of the five toes inclosed in the massive hoof of the
elephant or the webbed hind-paddle of the seal: and whether the hallux be the shortest of the five or the longest, it
has always the same number of phalanges whenever it is
present, provided it supports a nail, a hoof or a claw, in
the mammalian series.

The satisfaction felt by the rightly constituted mind must ever be great in recognising the fitness of parts for their appropriate functions ; but when this fitness is gained, as in the great-toe of the foot of man and the ostrich, by a structure which at the same time betokens harmonious concord with a common type, the prescient operation of the One Cause of all organization becomes strikingly manifested to our limited intelligence.

It is interesting to perceive both in the human hand and foot that the digits that have been most modified either by excess or defect of development are precisely those that are the least constant in the mammalian series—the two, for example, that form the extremes of the series ; whilst the three intermediate digits are more conformably and equably developed.   In the hand, the ' digitus medius '—the most constant of all in the vertebrate series, and most entitled to be viewed as the persistent representative of the terminal segments of the primitive elementary ray,—still shows a slight superiority of size ; though few would be led thereby to suspect that the bones forming the three joints of this finger answer to those called ' great pastern-bone,' ' little pastern-bone,' and ' coffin-bone ' in the horse, and that the nail of this finger represents the hoof in the horse.

In the human foot the three more constant toes, *ii, iii, iv,* maintain more equality of size than their homotypes in the hand : the middle toe here also is the representative of the chief part of the hind-foot of the horse : but the fourth toe answers to that which, by excess of growth, becomes the chief member of the long and strong hind-foot of the kangaroo.   These and the like relations to the Vertebrate archetype, which, together with the principle of the fitness of things, govern the forms and proportions of parts of the human frame, cannot but be both interesting and useful to the artist, as being calculated to call his attention to differential characters, which, though constant, may be so slight as to escape attention until their true significance is made known.

The few examples of unmutilated feet from the works of the ancient Greek sculptors show, indeed, how truly their just observation of nature supplied the insight into the archetypal law, and guided them to an exact and beautiful indication of the affinities of the three middle toes as contrasted with the first and fifth, the distinctive characters of the last being as truly given as those of the great-toe *.

If we pause to take a retrospect of the ground over which we have been travelling, and consider the numerous and beautiful evidences of unity of plan which the structures of the locomotive members have disclosed,— evidences so little to be expected, à priori, seeing the different shapes and sizes of instruments adapted to such diversity of functions ;—when also we find that besides the general conformity of structure in the limbs of different species, a more special parallelism could be traced between the fore- and hind-limbs of the same species, no matter to what diversity of office they might be severally adapted—a parallelism or 'serial homology' demonstrable even to each little carpal and tarsal bone, from man down to the monodactyle horse,—the thinking mind cannot but be forcibly struck by such facts, and be impelled with the desire to penetrate further, and ascend if possible to the higher law or generalization from which those harmonies flow.

I think it will be obvious that the principle of final adaptation fails to satisfy all the conditions of the problem. That every segment and almost every bone which is present in the human hand and arm should exist in the fin of

* I have elsewhere cited examples from some great painters in which these characters have been overlooked, and the toes drawn " small by degrees and beautifully less " from the second to the fifth : the natural proportions given to the feet of the dead Saviour by the truthful and severe Francia, and by Sebastian del Piombo to those of the figures in the " Raising of Lazarus," contrast favourably with the conventional feet in some of the paintings by Correggio and Guido in our National Gallery.

the whale, solely because it is assumed that they were required in such number and collocation for the support and
movements of that undivided and inflexible paddle, squares
as little with our idea of the simplest mode of effecting the
purpose, as the reason which might be assigned for the
great number of bones in the cranium of the chick, viz. to
allow of the safe compression of the brain-case during the
act of exclusion, squares with the requirements of that act.
Such a final purpose is indeed readily perceived and admitted in regard to the multiplied points of ossification of
the skull of the human fœtus, and their relation to safe
parturition.   But when we find that the same ossific centres are established, and in similar order, in the skull of the
embryo kangaroo, which is born when an inch in length,
and in that of the callow bird that breaks the brittle egg,
we feel the truth of Bacon's comparison of ' final causes'
to the Vestal Virgins, and perceive that they would be
barren and unproductive of the fruits we are labouring to
attain, and would yield us no clue to the comprehension of
that law of conformity of which we are in quest.   And so,
again, with regard to the structural correspondences manifested in the locomotive members; if the principle of
special adaptation fails to explain them, and we reject the
idea that these correspondences are manifestations of some
archetypal exemplar on which it has pleased the Creator
to frame certain of his living creatures, there remains only
the alternative that the organic atoms have concurred fortuitously to produce such harmony*.

But from this Epicurean slough of despond every
healthy mind naturally recoils : and reverting therefore to
the hypothesis of the dependence of special and serial homologies upon some wider principle of conformity, we have
next to inquire, what is the archetype or essential nature
of the limbs ?

* 'Απὸ τῶν ἀτόμων σωμάτων ἀπρονόητον καὶ τυχαίαν ἐχόντων τὴν κί-
νησιν.—Epicurus.

Cuvier seems to have regarded the science of Comparative Anatomy to be not sufficiently advanced to enter upon this analysis with any prospect of success.

Oken's idea of the essential nature of the arms and legs is, that they are no other than 'liberated ribs'; "Freye Bewegungsorgane können nichts anderes als frey gewordene Rippen seyn*."

Carus, in his ingenious endeavours to gain a view of the primary homologies of the locomotive members, sees in their several joints repetitions of vertebral bodies (*tertiarwirbel*)—vertebræ of the third degree†—a result of an ultimate analysis of a skeleton pushed to the extent of the term 'vertebra' being made to signify little more than what an ordinary anatomist would call a 'bone.' But these transcendental analyses sublime all differences, and definite knowledge of a part evaporates in an unwarrantable extension of the meaning of terms.

I believe, however, that I have satisfactorily demonstrated that a vertebra is a natural group of bones, that it may be recognised as a primary division or segment of the endoskeleton, and that the parts of that group are definable and recognizable under all their teleological modifications, their essential relations and characters appearing through every adaptive mask ‡.

Subjoined is a view of such a natural segment as it exists in the thorax of a bird. c is the 'centrum,' which was originally a separate element: *n*, one of the walls of the canal for the 'spinal marrow' (myelon or trunk-portion of the neural axis); it was originally distinct from the centrum and from its fellow of the opposite side; it is the 'neurapophysis': *pl* indicates another pair of elements—the 'pleurapophyses,' which here maintain their primitive

---

* Lehrbuch der Natur-Philosophie, p. 330, 8vo, 1843.

† Urtheilen des Knochen und Schalengerüstes, fol. 1828.

‡ 'On the Archetype of the Vertebrate Skeleton,' 8vo, 1848, pp. 80—102.

distinctness, and grow long in order to aid in encompass-
ing the dilated canal or cavity for the great vascular centres : so modified, these elements are called 'ribs' or vertebral ribs. The elements more constantly related to the

Fig. 7.

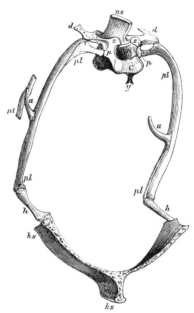

Natural skeleton-segment, 'osteocomma' or 'vertebra.' Thorax of
Bird.

protection of the vascular or hæmal axis are those marked
h, the 'hæmapophyses,' here displaced from the centrum
by the development of the heart and lungs and other organs of the embryonic vascular layer : they then get the
special name of 'sternal ribs,' or that of 'costal cartilages'
when their ossification is arrested. The hæmal arch is
here completed by an unusually expanded 'hæmal spine,'
hs, which coalesces with those of contiguous segments, and
they are then collectively denominated the 'sternum.' The

neural arch is completed by an answerable piece above, *n s,* which here retains its normal size and form as a 'neural spine.' But, where the neural canal is expanded for great developments of the nervous centres, the brain, *e. g.,* analogous to the development of the vascular centres in the 'thorax,' the neural canal is correspondingly expanded and forms a 'cranium': other elements besides the 'neurapophyses' enter into the formation of its walls, and the 'neural spine' is expanded horizontally, often retains its individuality, and, like the 'sternum' in the thorax of the bird, receives a special name, *e. g.* 'parietal' or 'frontal.' Two other distinct elements, *a, a,* in fig. 7, are attached at one end to the hæmal arch, and project backwards, overlapping, in the bird, the succeeding arch. Less constant, secondary or derivative processes more or less complicate the typical segment, but are not essential to it: of these are shown in fig. 7, *zz* the 'zygapophyses,' *pp* the 'parapophyses,' *dd* the 'diapophyses,' and *y,* the 'hypapophysis.'

As the expanded nervous centres contract to a column or trunk of moderate and pretty uniform diameter, so also do the vascular centres; and where the vascular axis is reduced to the dimensions of the nervous axis above, the

Fig. 8.

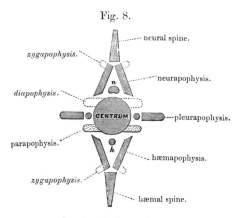

zygapophysis. ---- neural spine.

----neurapophysis.

diapophysis.

CENTRUM ----pleurapophysis.

parapophysis.

--- hæmapophysis.

zygapophysis.

--- hæmal spine.

Ideal typical vertebra.

vertebra resumes a more symmetrical character, by the corresponding reduction of the hæmal arch : in the crocodile, *e. g.* this is formed by the hæmapophyses which ascend to contact with the centrum ; the pleurapophyses being shortened, projecting outwards, and anchylosed as processes : but such a vertebra, when analysed as it is developed, resolves itself very nearly into the ideal type given in the diagram fig. 8.

n is the neural axis or myelon, and *h* the hæmal axis or aorta, protected by their respective arches. The Roman type is used for the 'autogenous elements,' or those usually developed from distinct centres of ossification ; and the italics denote the parts more properly called 'processes' which shoot out from the preceding elements.

On comparing this form of the segment with the foregoing one, it will be seen that they differ by altered proportions with changed positions of certain elements ; but it is important to our present inquiry to notice that in figure 7 there is an additional bone marked *a* which projects as an appendage from each side of the hæmal arch.

As we find such appendages in the thoracic segments of the crocodile (Pl. I. fig. 3, *a, a*), and in the corresponding abdominal segments of the skeleton of most bony fishes (*ib.* fig. 2, *a, a*), where they are usually longer, and extend through the flesh to the skin, we may regard them also as parts of the primitive segment or vertebra, though less constant than the arches that support them.

The attempt to decipher the essential nature of limbs will lead us in the first place to their comparison with one or other of the elements of the typical segment. And, having arrived at the demonstration of such segment, and proved that every modification of the axis of the skeleton has its seat in one or other of the segmental or 'vertebral' elements, we may enter upon this track of inquiry with hopeful confidence in the results.

I have already observed that Oken selected the rib as

the seat of those developments from which the varied forms of the limbs of animals result : the truth of this idea will be tested by tracing the progressive simplification of the limb in the Vertebrate series.

We already find it in the Mammalian series reduced almost to one long jointed ray in the horse (fig. 4), two other terminal forks being feebly indicated by the little splint-bones. In the lower Reptilia we find the limbs progressively diminish in relative bulk to the body and to their sustaining arch. In the amphiuma (Pl. I. fig. 8) each limb is a jointed appendage, consisting first of a single segment (53), then of a bifid segment (54 & 55), lastly of a bifid (57) or a trifid segment (*A. tridactylum*) ; in these segments we have the homologues respectively of the humerus or femur, of the two bones of the fore-arm or leg, and of the three middle and most constant toes. In the proteus (fig. 10) the terminal forks are diminished to two on the hind-limb, and in the *Amphiuma didactylum* to two digits on both fore- and hind-limbs. In the apteryx (fig. 9) the second segment of the stunted fore-limb is bifid, but the last segment is simple or monodactyle. One other step in the series, and, in the lepidosiren, we find the limb reduced to an unbranched ray, modified only by a vegetatively multiplied succession of simple segments (Pl. I. figs. 7 & 9).

But in none of these instances do the limbs diverge, like the free or false ribs in the parachute of the little flying dragon, from the vertebral centrums or neural arches.

Under whatever grade of structure the limbs exist, they are supported by inverted arches, the anterior being the ' scapular arch,' the posterior the ' pelvic arch ' : and these arches remain when every trace of limb has disappeared ; as for example, the scapular arch in the *Anguis* and *Muræna*, and a rudiment of the pelvic arch in the Cetacea.

The question, therefore, which we have now to determine

is, what is the nature or general homology of these arches? and first of the scapular arch?

Of the special homology of its parts, to the extent that we have traced them, there exists I believe no difference of opinion amongst anatomists.

The scapula is usually as broad and flat in mammals as the first rib of the whale is; it becomes narrow and vertical in the ornithorhynchus; continues long and slender in birds and saurians; is a columnar rib-like bone in the Chelonia, where it is as straight as the first rib in the ox; and it retains the same form in the amphiuma (Pl. I. fig. 8, 51) and lepidosiren (*ib.* fig. 7, 51).

The scapula is the upper element or pier of an inverted arch, completed, with a few exceptions, by a second elongated bone, converging towards its fellow on the opposite side, and joining it either directly or through the medium of a single interposed key-bone, called 'sternum,' which bone answers below to the key-bone or 'spine' of the neural arch above. The special name of the second bone is the 'coracoid.' The crocodile (Pl. I. fig. 3) affords a good example of the proper scapular arch so composed, *i. e.* consisting of a pair of scapulæ (51), a pair of coracoids (52), and the sternum or episternum (52′).

In all reptiles and in all birds the arch is thus completed: in fishes the coracoids meet without the interposition of a sternum (Pl. I. figs. 2 & 7, 52). In most oviparous vertebrates the arch is complicated by a second inferior element, parallel or nearly so with the coracoid; behind it in fishes (fig. 2, 58), before it in reptiles, birds (fig. 4, 58) and monotremes: this accessory bone is the 'clavicle': when it unites below with its fellow, as in birds, it is called the 'merry-thought' (*os furcatorium*); but it is wholly wanting in some species, as in certain ground parrots and the apteryx (fig. 9). The clavicle is wanting altogether in entire groups of Mammalia, *e. g.* in the Cetacea, and in all the Hoofed orders. But the coracoid is always

present, although its complete development, which is the
rule in the lower classes of Vertebrata, becomes the exception in Mammalia, where it reaches the sternum only in
the ovoviviparous ornithorhynchus and echidna. In the
rest of the class the coracoid projects as a longer or shorter
process from the scapula, but differs from the other 'processes' of that bone in having a separate centre of ossification; it is, in short, an autogenous element, whilst the
'spine' and 'acromion' are merely exogenous growths.

Fig. 9.

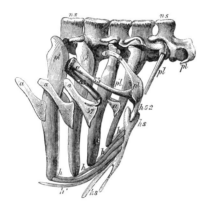

Part of the thorax with the scapular arch and appendage. *Apteryx
australis.*

In many Mammalia the clavicle appears to take the place
of the fully-developed coracoid of the Ovipara, and com-
pletes the scapular arch by extending from the acromion
to the sternum, as *e.g.* in our own skeleton (Frontispiece
and Pl. I. fig. 6, 58*). But it will be readily understood by
what has been stated with regard to the relative constancy,
both as to connexion and existence, of the coracoid, that
this is the true inferior element of the scapular arch, and

* A front view of the human scapular arch is given at p. 46 of Sir
C. Bell's 'Bridgewater Treatise.'

that its place is, so to speak, usurped by the clavicle in man, just as the chain of suborbital bones in the skull of the fish takes the place of the true zygomatic arch of higher Vertebrates*.

Leaving, however, for the present the discussion of the general homology of the clavicle, and returning to that of the scapular arch, as it exists in its integrity and simplicity, in the apteryx or crocodile, the question before us is, with what part of the typical segment of the vertebral skeleton can it best be compared? The answer is at once suggested by the hæmal arch, as it exists in the thoracic region of any air-breathing vertebrate, in which the parallel parts are obviously presented in similar connexion with each other. The scapula (Pl. I. fig. 7, *pl*) answers to the pleurapophysis or vertebral rib (p. 42, fig. 7, *pl*), the coracoid (*h*) to the hæmapophysis or sternal rib (*h*), and the scapular arch is completed in most vertebrates by a bone of the sternum or hæmal spine (Pl. I. fig. 3, 52′). In harmony with this determination is the interesting correspondence in the range of its diversities of form and proportion which the scapula presents with that which may be traced in the thoracic pleurapophyses or 'vertebral ribs.' I have already cited the whale, *e. g.* as manifesting such a ' rib ' with proportions as broad and flat as any mammalian scapula ; and numerous instances will occur to the anatomist of 'ribs' that are as long, narrow, compressed and gently curved as is the scapula in the bird ; or that are straight and columnar like the scapula in the tortoise.

Any difficulty of appreciating or hesitation in accepting such general homological views by the anthropotomist arises, he may be assured, from his habitual and exclusive contemplation of the vertebral element in question under one only of its morphological phases, and that usually an extreme and an exceptional one. Even the steps in the progressive degradation of the true scapular arch which we

* On the Archetype of the Vertebrate Skeleton, p. 62.

witness in the mammalian class are closely paralleled by
the ordinary costal arches: thus the twelfth vertebral rib
in man, with its sternal rib or cartilage curtailed and appended to its free extremity, as a process, repeats the
condition of the scapula with its shortened coracoid appended and anchylosed to it, as a process, in man also, and
in most mammals.

The posterior costal arch in the bird's thorax, to the
sternal rib (hæmapophysis) of which may be attached the
same element of a contiguous arch, unconnected with its
proper vertebral rib, as seen in the apteryx (fig. 9, *h*, *h'*),
parallels that condition of the scapular arch where it is
complicated by an appended clavicle,—the hæmapophysis
of a second inverted arch.

The scapula, however, differs, in most of the higher
Vertebrata, from the thoracic ribs in being more or less
turned from the perpendicular to the horizontal direction: this is particularly the case in birds, where it lies
almost at right angles across those ribs, as in fig. 4,
Pl. I. Its progressive assumption, however, of the vertical position, or one more or less parallel with that of
the ordinary ribs, as the species descend in the scale, would
have much weight, in forming a judgment as to the essential nature of the scapula, with the physiologist who appreciated the law that the Archetype is progressively departed from as the organization is more and more modified
in adaptation to higher and more varied powers and actions.
But what is still more significative of the nature of the
scapula is the discovery that its vertical position in reptiles
and fishes is a retention of that position which the bone
manifests on its first appearance in all Vertebrata. The rotation of the scapula from its primitive verticality to whatever approach to the horizontal line, or axis of the vertebral bodies, the exigencies of the full-grown animal may
require, its primitive appearance close to the occiput, and
its longer retention of a place anterior to the first thoracic

rib, are developmental phænomena which cease to be mere empirical facts*, but receive their explanation and become intelligible by the recognition of the general homology of the scapula which I am now attempting to illustrate.

Sufficient, I trust, has been adduced to show that, of all the elements of the typical vertebra or primary segment, the scapula can be compared only with the 'pleurapophysis,' and the coracoid only with the 'hæmapophysis.' But it may be objected, that the ordinary costal or hæmal arch has been detached from its centrum for the purpose of this comparison. True! And the scapular arch in mammals, birds and reptiles is a hæmal arch so dislocated,—a statement which I do not hesitate to make under a pledge to demonstrate the proper centrum and the rest of the segment or vertebra to which it belongs.

In the first place I may remark, that the dislocation of parts of admitted costal or hæmal arches is no new thing. Why does the human anatomist refer the fourth costal arch to the fourth dorsal vertebra and not to the fifth, to the centrum and neurapophyses of which it is equally connected? The head of the rib is applied half to one centrum half to the other: the upper border of the neck of the rib, viewed in the upright skeleton, articulates with the upper neural arch, the tubercle of the rib with the diapophysis of the lower neural arch: the human anatomist, in restoring this displaced hæmal arch to its proper segment, has availed himself of the same kind of inquiry and comparison† which has been instituted in order to detect the segment from which the scapular arch has strayed. The same kind of comparison was equally requisite in order to determine

---

* See Rathké, 'Ueber die Entwickelung der Schildkröten,' 4to, 1848, pp. 182, 183. The different position of the scapula in fig. 9, 51, from that in the typical Bird's skeleton (Pl. I. fig. 4, 53), is due to a retention of an embryonic phase in the apteryx.

† See 'On the Archetype of the Vertebrate Skeleton,' pp. 118, 119.

the true segments to which the displaced neural arches in the sacrum of the bird belonged*.

But here let us examine how far down the vertebrate scale the scapular arch continues to manifest itself as a hæmal arch displaced: and let us ascertain whether Nature affords us any instance of such a retention of the ideal type as would be exemplified by the connection of that hæmal arch with its proper centrum and neural arch? The object we are in quest of is soon attained.

The desired instance is mostly clearly and satisfactorily afforded by the eel-like fish to which I have already referred as affording the most simple type of locomotive extremity, I mean the lepidosiren. We see that the upper element of the arch is now articulated to the neural arch of the occipital vertebra: and if we remove this segment of the skull and take a full view of it, as in fig. 7, Pl. I., we have a vertebra which closely conforms to the typical one illustrated by the thoracic segment of the bird, p. 42, fig. 7. In both figures, c is the centrum, n the neurapophyses, s the neural spine, pl the pleurapophyses, and h the hæmapophyses: the hæmal spine alone is wanting to complete the parallel: and we know that it is not more essential to the composition of the hæmal arch, than the neural spine is to the completion of the neural arch.

There can be no doubt that the bone which represents the rib or hæmapophysis in the occipital segment of the amphibious fish (Pl. I. fig. 7) is the bone which receives the special name of ' scapula' in higher animals. We have traced it under all its modifications down to the low batrachian reptile (Pl. I. fig. 8), in which its special homology was clearly recognised by Cuvier; and between the straight columnar upper element (51) of the inverted arch which supports the simple pectoral limb of the amphiuma, and the upper element (51) of the inverted arch which supports the still more simple limb in the lepidosiren, the resem-

* 'On the Archetype of the Vertebrate Skeleton,' pp. 118, 159.

blance is much closer than that between the scapula of man and the scapula of the mole, or between this and the scapula of the ornithorhynchus.

Since also comparative anatomists have arrived at one conclusion in respect to the homology of the stunted coracoid in man with the coracoid element that stretches from scapula to sternum in the ornithorhynchus and all airbreathing Ovipara, it would be captious to deny the same relationship between the coracoid of the amphiuma (h 52) and that of the lepidosiren (h 52). The sole fact lending colour to such negation is the slight difference in the place of attachment of the diverging appendage, which is wholly in the coracoid in the lepidosiren, instead of partially, as in the amphiuma,—a difference which, with that in the upper connections of the scapula, is characteristic of their respective classes.

The main fact, then, is established, viz. that the inverted arch (51, 52, fig. 7, Pl. I.) which supports the pectoral ray in the lepidosiren is the scapular arch, and Nature here shows us its attachment to the occiput (c 1, n 2, s 3) in the relation of the hæmal arch of the occipital vertebra.

Nor is the lepidosiren an exceptional instance of this connection and relation. It forms but one of a vast class of Vertebrata—by far the most numerous and widely-dispersed class—which manifest the same attachment of the scapular arch. The pectoral fin (Pl. I. fig. 2, 54–57, a) which consists, in most fishes, of a multiplication of jointed rays like that of the lepidosiren, is supported by an inverted bony arch, attached to the neural arch of the occiput, and completing that segment of the skeleton by forming its hæmal arch*. Whatever be the size, form or function of the pectoral limb, its supporting arch maintains this position and connection in all osseous and some cartilaginous (Chimæroids, Sturionidæ) fishes: and here, also, we find

---

* 'Archetype of the Vertebrate Skeleton,' p. 107.

the arch in the *Muræna* and *Gymnothorax,* in which every trace of its appendage, the pectoral fin, has disappeared.

How then is this connection of the scapular arch to be interpreted? If we were prepossessed by the habitual contemplation of its detached state in man and the higher Vertebrata, we might deem such occipital connection an anomaly, an 'instantia devians,' and so in fact it has hitherto been viewed by those anatomists who have entered into the study of the higher generalizations of their science. Geoffroy St. Hilaire, one of the boldest speculators in the mine of transcendental ideas, calls it in his 'Anatomie Philosophique' (p. 481) 'une amalgame inattendue:' and in one of his Memoirs speaks of it as " Disposition véritablement très singulière, et que le manque absolu de cou, et une combinaison des pièces du sternum avec celles de la tête pouvoient seuls rendre possible *."

We no longer, however, believe that the vertebræ of the neck are absolutely wanting in fishes; but recognize them with their pleurapophyses more normally or typically developed than in the corresponding region of the crocodile. Homotypes of the thoracic sternal bones are doubtless present in the head of fishes, as in that of other Vertebrates, where they are exemplified by the basihyal (see 41, figs. 2 to 6, Pl. I.); but it is not requisite to conclude that the median bones (41, 42, 43, fig. 2) of the hyoid arch, or the analogous ossicles of the transitory splanchnic branchial arches, in fishes, are the homologues of 59 and 60 in figs. 3 to 6, Pl. I., in order to arrive at the comprehension of the connection of the scapular arch with the occipital vertebra in the lepidosiren and the majority of its class.

Viewing that vertebra, like the other natural segments of the skeleton, to be conformable to the type as illustrated by fig. 7, p. 42, it would be incomplete without the hæmal arch formed by the scapula and coracoid. And further, there

* Annales du Muséum, tom. ix. p. 361.

is no other arch which could so complete the occipital vertebra. There are three neural arches in advance of the
occipital one, in fig. 2, viz. the parietal (5, 6, 7, 8), the frontal
(9, 9', 10, 11, 12), and the nasal (13, 14, 15) ; and there are
three hæmal arches, viz. the maxillary (20, 21, 22), the
mandibular (28, 29, 32), and the hyoid (38, 39, 40, 41). With
the exception of the hyoid arch, which is slightly depressed, each of these hæmal arches is directly connected
with its respective neural arch, just as the scapular arch is
connected with the occipital : there is no other arch to
supply the place of the hæmal one of the occipital segment, if the scapular arch be removed.

Are we then to view the instances of its detachment
from its piscine connections as exemplifying the normal
or typical conditions of such arch ? Surely not ; as well
might we consider the displaced hæmal arches of the
human thorax to manifest their typical positions and
connections, and regard their direct and exclusive connections with their proper centrums in reptiles and fishes as
the exception and the anomaly. But that which has happened to the crocodile in the restoration of the ribs, homologous to the second and tenth inclusive of man, each
from the interspaces of two vertebræ to the body and neural
arch of one vertebra, i. e. to the connections which the
first, eleventh and twelfth ribs retain in the Human skeleton, has likewise taken place in fishes in regard to the modified rib forming the scapula. So true is the Baconian
aphorism regarding the power of interpretation consequent
upon the knowledge of the archetype or ' via Naturæ !'
" Inter prærogativas instantiarum ponemus loco octavo *in-
stantias deviantes*; errores scilicet naturæ, et vaga ac mon-
stra : ubi natura declinat et deflectit a cursu ordinario.
Differunt enim errores naturæ ab instantiis monodicis in
hoc, quod monodicæ sint miracula specierum, at errores sint
miracula individuorum. Similis autem fere sunt usus ; quia
rectificant intellectum adversus consueta, et revelant formas

communes. Neque enim in his etiam desistendum ab inqui-
sitione, donec inveniatur causa hujusmodi declinationis.
Veruntamen causa illa non exsurgit ad formam aliquam
proprie, sed tantum ad latentem processum ad formam.
Qui enim vias naturæ noverit, is deviationes etiam facilius
observabit. At rursus, qui deviationes noverit is accuratius
vias describet *."

Most of the mistakes in the attempts to ascertain the
typical or essential nature of parts of the skeleton, and
almost all the impediments and opposition to the prosecution of this main end of anatomical science, have arisen
from its study being confined to that by-path in which it
is usually commenced, viz. where the course of development has reached the highest form of animal life, in which
modification of each part in mutual subserviency to another is greatest, and the deviation from the archetype is
in the same degree. The rectification of the mistakes and
the acquisition of a more catholic feeling towards the study
are gained by pursuing the broader high-road of organic
nature, where those forms are offered to our contemplation in which vegetative uniformity most prevails, and the
archetype is least obscured by purposive adaptations.

If therefore we find in that class which best displays the
conditions for solving the problem immediately before us,
that the connections of the scapular arch are such as to
complete a typical segment, which otherwise would be abnormal by defect, we must conclude that the type is here
adhered to ; and that, although these connections are abrogated in all the other Vertebrate classes, they. nevertheless,
are the ' instantiæ deviantes,' and are exceptions in regard
to the rule of the archetype, notwithstanding the actual
numerical superiority of the instances.

And the latter fact leads us to another consideration.
This superiority was not always such as it now is. Time

---

* Novum Organum Scientiarum, lib. ii. Aph. xxix.

was when fishes were the sole representatives of the Vertebrate subkingdom in this planet. During the long periods
antecedent to the formation of the coal-measures, the vertebrate type was exclusively manifested by forms in the great
majority of which the scapular arch was articulated to the
occiput. Subsequent changes in our planet have height-
ened and varied the conditions of animal existence, but the
primitive pattern of the skeleton may be discerned beneath
all the superinduced modifications.

We perceive a return to it, as it were, in the early phases
of development of the highest organized of the actually existing species, or we ought rather to say, that development
starts from the old point; and thus, in regard to the scapula, we can explain the constancy of its first appearance
close to the head, whether in the human embryo or in that
of the swan, and also its vertical position to the axis of the
spinal column, by its general homology as the rib or ' pleurapophysis ' of the occipital vertebra.

We observe, as might naturally be expected, that its
degree of displacement is least in those air-breathing Vertebrata that make the nearest approach to fishes : whether
in general structure, as e. g. the siren and amphiuma, or
in outward form, as e. g. the Cetacea. In comparing the
crocodile's skeleton (fig. 3. Pl. I.) with that of the fish, the
chief modification that distinguishes the occipital segment
from its homologue in the fish, is the absence of its attached hæmal arch. We recognise, however, the special
homologues of the chief piscine constituents of that arch in
51 and 52; but the upper or suprascapular piece (50) retains, in connection with the loss of its proximal or cranial
articulations, its cartilaginous state, and is not shown in
fig. 3: the scapula (51) is ossified, as is likewise the coracoid (52), the lower end of which is separated from its fellow by the interposition of a median, symmetrical, partially
ossified piece called ' episternum ' (hs). The power of recognising the special homologies of 50, 51, and 52 in the

crocodile, with the similarly numbered constituents of the arch H I in fishes (fig. 2), though masked not only by modifications of form and proportion but even of very substance, as in the case of 50, depends upon the circumstance of these bones constituting the same essential element of the archetypal skeleton : for although in the present instance there is superadded to the adaptive modifications above-cited the rarer one of altered connections, Cuvier does not hesitate to give the same names (suprascapulaire) to 50 and (scapulaire) to 51, in both fish and crocodile : but he did not perceive or admit that the narrower relations of special homology were a result of, and necessarily included in, the wider law of general homology. According to the view which I am attempting to establish and illustrate, we discern in 50 and 51 a teleologically compound *pleurapophysis*, in 52 a *hæmapophysis*, and in *hs* the *hæmal spine*, completing the hæmal arch.

The general relations of the scapulo-coracoid arch to a hæmal or costal one was recognised, as I have already observed, by Oken. This philosopher, having observed the free cervical ribs in a specimen of the *Lacerta apoda* (*Pseudopus*), deemed them representatives of the scapula, and this bone to be, in other animals, the coalesced homologues of the cervical pleurapophyses*. In no animal are the conditions for testing this question so favourable and obvious as in the crocodile (fig. 3. Pl. I.): not only do cervical ribs coexist with the scapulo-coracoid arch, but they are of unusual length and are developed from the atlas as well as from each succeeding cervical vertebra : we can

---

* "Auch die Scapula nicht *ein* Knochen, sondern wenigstens eine aus fünf Halsrippen zusammengeflossene Platte ist."—*Programm über die Bedeutung der Schädelknochen*, 4to, 1807, p. 16. He reproduces the same idea of the general homology of the scapula in the 'Lehrbuch der Natur-philosophie,' 1843, p. 331, ¶ 2381. Carus also regards the scapulo-coracoid arch as the reunion of several (at least three) protovertebral arches of the trunk-segments. 'Urtheilen des Knochen und Schalen gerustes,' fol. 1828.

also trace them beyond the thorax to the sacrum, and throughout a great part of the caudal region, as the sutures of the apparently long transverse processes of the coccygeal vertebræ demonstrate in the young animal; the lumbar pleurapophyses being manifested at the same period as cartilaginous appendages to the ends of the long diapophyses.

The scapulo-coracoid arch (51,52), both elements of which retain the form of strong and thick vertebral and sternal ribs in the crocodile, is applied in the skeleton of that animal over the anterior thoracic hæmal arches. Viewed as a more robust hæmal arch, it is obviously out of place in reference to the rest of its vertebral segment. If we seek to determine that segment by the mode in which we restore to their centrums the less displaced neural arches in the sacrum of the bird (fig. 10, $n1-n4$), we proceed to examine the vertebræ before and behind the displaced arch with the view to discover the one which needs it in order to be made typically complete. Finding no centrum and neural arch without its pleurapophyses from the scapula to the pelvis, we give up our search in that direction; and in the opposite direction we find no vertebra without its ribs until we reach the occiput: there we have centrum and neural arch, with coalesced parapophyses—the elements answering to those included in the arch fig. 7, c, n, p, ns, but without the inverted arch pl, h, hs; which arch can only be supplied, without destroying the typical completeness of antecedent cranial segments, by a restoration of the bones 51–52, fig. 3, Pl. I. to the place which they naturally occupy in the skeleton of the fish (fig. 2). And since anatomists are generally agreed to regard the bones 51–52 in the crocodile (Pl. I. fig. 3) as specially homologous with those so numbered in the fish (*ib.* fig. 2), we must conclude that they are likewise homologous in a higher sense; that in fig. 2 the scapulo-coracoid arch is in its natural or typical position, whereas in the crocodile it has been displaced for a special purpose. Thus, agreeably with a general

principle, we perceive that as the lower vertebrate animal illustrates the closer adhesion to the archetype by the natural
articulation of the scapulo-coracoid arch to the occiput, so
the higher vertebrate manifests the superior influence of
the antagonising power of adaptive modification by the removal of that arch from its proper segment.

The scapula retains the more common cylindrical long
and slender rib-like form of the pleurapophysis in the chelonian reptiles, where, from the greater length of the neck,
it has retrograded further than in the crocodile from its
proper centrum, and is placed not upon, but within, an
anterior thoracic hæmal arch, by virtue of the great expansion to which the pleurapophysis of that arch has been
subject. Here the rib has become a broad and flat bone*,
whilst the scapula has retained its primitive rib-like shape.

If the arguments founded upon the relations of the scapulo-coracoid arch to the segments of the skeleton in osseous fishes and crocodilians be admitted to sustain the conclusion here drawn from them, that arch must be held to
form the hæmal complement of the occipital vertebra in all
animals.

Bojanus, in illustrating his vertebral theory of the skull
by the osteology of the *Emys Europæa*, thus defines the

"VERTEBRA OCCIPITALIS, SIVE CAPITIS PRIMA.

" Basis occipitis, seu *corpus* hujus vertebræ,

" Pars lateralis occipitis, sive *arcus*,

" Crista occipitalis, *processus spinosi* loco,

" Cornu majus hyoidis, *costæ vertebræ* occipitalis compa-
randum †."

He adds a dotted outline of the hyoid arch to complete
the *vertebra occipitalis*, in tab. xii. fig. 32, B. 1 of his
beautiful Monograph.

---

* By what process this metamorphosis is effected I have explained
in a Memoir read before the Royal Society, January 18th, 1849.

† Anatome Testudinis Europææ, fol. 1819, p. 44.

Supposing the special homology of the middle cornua of the hyoid of the chelonian, so represented and compared to ribs by Bojanus, with the stylo-, epi- and cerato-hyals of the fish (Pl. I. fig. 2, 38, 39, 40) to have been correct, which the metamorphoses of the hyoid and branchial arches in the batrachians disprove, the singular and highly interesting change of position as well as shape of the true cerato-hyals, during the same metamorphosis, prepares us to expect a retrogradation of the hyoid arch in respect to its proper centrum, in the skulls of the air-breathing vertebrates. In the young tadpole the thick cartilaginous hyoidean arch* is suspended, as in fishes, from the tympanic pedicle: the slender hyoidean arch of the mature frog is suspended from the petrosal capsule†. The mandibular arch has, also, receded; and the scapular arch, which, at its first appearance, was in close connection with the occiput, further retrogrades in the progress of the metamorphosis to the place where we find it in the skeleton of the adult frog.

The argument, therefore, may be summed up as follows. The position of the pleurapophyses in the human thorax and that of most mammals; and the position of the neurapophyses in the dorsal vertebræ of chelonians and in the sacral vertebræ of dinosaurians and birds, show that a change of relative position in respect of other elements of the same vertebra may be one of the adaptive modifications to which even the most constant and important of those elements are subject. Instead of viewing such shifted arches as primary and independent constituents of the skeleton, we recognise them as secondary and derivative parts of a natural segment or whole, and we trace their relation to the stationary elements—the centrums of the primary vertebral segments.

* Cuvier, Ossem. Foss. v. pt. ii. pl. 24, fig. 23 a.

† Ib. fig. 27 a:—an intermediate stage is shown at fig. 25. Dugés and Reichert confirm and further illustrate this change of position of the hyoidean arch.

Thus, commencing, for example, with the anterior of the sacral vertebræ of the ostrich, A in fig. 10, we observe that,

Fig. 10.

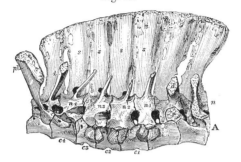

Seven sacral vertebræ of a young ostrich (*Struthio camelus*).

besides supporting its own neural arch *n*, it bears a small portion of that of the next vertebra : the third neural arch (*n* 1) has encroached further upon the centrum of the vertebra in advance ; and thus, in respect to the neural arch (*n* 2), if it were viewed with the centrums, *c* 2 and *c* 1, upon which it equally rests, apart from the rest of the sacrum, it would appear to appertain equally to either, and be referable to the one in preference to the other quite gratuitously, to all appearance. Nevertheless *n* 2 is proved, by the intermediate changes in antecedent neural arches, to belong actually, and in no merely ideal or transcendental sense, to *c* 2 altogether, and not to the segment of which *c* 1 is the centrum ; and in tracing the modifications of those sacral vertebræ which follow *c* 2, we find *n* 4 to have regained nearly the whole of its centrum, *c* 4, and the normal relations of the elements are quite restored in the succeeding vertebra.

Now let us suppose the habits of the species to have required a more extensive displacement of the arch (*n* 2) and its appendages : if its formal characters as a neural arch were still retained beneath the adaptive development superadded to the adaptive dislocation, and if the segments before and behind the centrum $c\,2$ were found complete, and that centrum alone wanting its neural arch; would the mere degree of modification in respect of relative position nullify the conclusion that the shifted arch appertained to such incomplete segment, and forbid that restoration to the typical condition, which no anatomist, it is presumed, will dispute in the case of $n\,2$, $c\,2$, fig. 10?

The anthropotomist, by his mode of counting and defining the dorsal vertebræ and ribs, admits, unconsciously perhaps, the important principle in general homology for which we are contending, and which, pursued to its legitimate consequences and further applied, demonstrates that the scapula is the modified rib of that centrum and neural arch which he calls the 'occipital bone,' and that the change of place which chiefly masks that relation (for a very elementary acquaintance with comparative anatomy shows how little mere form and proportion affect the homological characters of bones) differs only in extent and not in kind from the modification which makes a minor amount of comparative observation requisite in order to determine the relation of the shifted dorsal rib to its proper centrum.

With reference, therefore, to the occipital vertebra of the crocodile, if the comparatively well-developed and permanently distinct ribs of all the cervical vertebræ prove the scapular arch to belong to none of those segments, and, if it be wanting to complete the occipital segment, which it actually does complete in fishes, then the same conclusion must apply to the same arch in other animals, and we must regard the occipital vertebra of the tortoise as completed below by its scapulo-coracoid arch, and not, as Bojanus supposed, by its hyoidean arch *.

* Geoffroy St. Hilaire selected the opercular and subopercular bones to form the inverted arch of his seventh (occipital) cranial vertebra ('Tableau de la Composition de la Tête osseuse de l'Homme et des Animaux,' cited in Cuvier's 'Histoire de Poissons,' tom. i. p. 230. See

The facts and arguments that determine the nature of the hæmal arch of the occipital vertebra in reptiles equally apply to that in the bird (see fig. 4, 51, 52, Pl. I.). The extent of its displacement, it is true, has been greater : not seven only, but seven and twenty vertebræ may intervene between its actual position and that of the rest of its proper segment. But this difference of extent of displacement ought no more to hide the true relationship of the scapulo-coracoid arch to its proper position and typical segment in the skeleton, than the jugular position of the ventral fins of a fish (Pl. I. fig. 2, v'') prevents the ichthyologist from determining their nature and consequently their proper position in relation to the archetype skeleton.

In Mammalia we find the scapula retaining its primitive vertical position and much of its elongated narrow form in the low ovoviviparous monotremes ; and the arch in them is completed by the coracoid. We observe the scapula long and slender in the mole (fig. 2, 51) ; but in most of the class it is developed into a broad quadrate or triangular plate, with outstanding exogenous processes called 'spine' and 'acromion.' The hæmapophysial element is also reduced to an appendage, as in the false ribs at the back part of the thorax, and coalesces as a process (Pl. I. fig. 4, 52) with the pleurapophysis (51). In many mammals and in man, the arch is completed by the pair of bones called clavicles (Pl. I. fig. 6, 58) which coexist with the coracoids in most birds and reptiles. I have elsewhere adduced the facts and arguments which show that the bone 58, figs. 4, 5 & 6, Pl. I., is the special homologue of the bone 58 in the fish (fig. 2, 51 & 52) * : and its anterior position to the coracoid in the air-breathing Vertebrata is no valid

also the Table III. and note 11 in my Work 'On the Archetype,' &c. p. 172), and took no account of the instructive natural connections and relative position of the hyoidean and scapular arches in fishes.

* 'On the Archetype and Homologies of the Vertebrate Skeleton,' pp. 19, 99, 133.

argument against the determination, since in these we have shown that the true scapular arch is displaced backwards. The parallelism of the clavicles with the coracoids, and their inferior convergence, indicate their serial homology with the coracoids, and consequently their general homology as ' hæmapophyses '; and I regard the clavicle in its relations to the vertebrate archetype (Pl. I. fig. 1), as the displaced hæmapophysial element of the atlas, to which segment its true relative position is shown in the same low organized class (fig. 2, 58) in which the typical position of the scapular arch is likewise retained.

Having thus traced out and determined the nature, 'bedeutung' or general homology of the scapular arch, the next step, in regard to the limb appended to it, is plain and easy. We saw that the typical segment of the vertebrate skeleton (p. 42, fig. 7) was occasionally complicated by an appendage (a) developed from each side of the hæmal arch, serving to attach that arch to the next in succession in the thorax of birds (a, p); but diverging into the flesh and adding to the leverage of the locomotive powers in the abdomen of fishes (Pl. I. fig. 2, a, a). The pectoral extremities exhibit varied developments of this appendage; but, if our determination of their general homology be the true one, we ought to trace them under progressive stages of reduction and simplification to their primitive or archetypal character.

Already, in the class of Birds, we find the diverging appendages of the scapulo-coracoid arch of the apteryx (fig. 9, 53–57, a) reduced in bulk nearly to the dimensions of those of the thoracic-costal arches (a, a), but projecting into the flesh like the costal appendages of fishes; and, being a longer ray, protruding its extremity beyond the tegumentary surface, but to so short an extent that the whole of this monodactyle limb or rudimental wing is concealed by the plumage.

The proportion of the fore-limb to the supporting arch

in the proteus and amphiuma (Pl. I. fig. 8, 53–57, *a*) is hardly greater than that of the ray-like appendage to the abdominal rib in many fishes.

In the lepidosiren (*ib.* fig. 7, 53–57, *a*) we actually find the diverging appendage of the scapular arch retaining its elementary form of the ray, and differing only by its segmentation and relative length from its serial homologues (*ib.* fig. 2, *a. a*) in the succeeding segments of the skeleton of better ossified fishes. But by virtue of its elongation it extends beyond the surface of the body, and carries with it a sheath of the integument.

In the view of the Archetype skeleton (Pl. I. fig. 1), the pectoral limb (53–57, *a*) is represented under this form as the diverging appendage of the fourth or occipital hæmal arch (50, 51, 52, 52′), and its serial homology with the shorter appendages (*a, a*) of the succeeding arches is unmistakeable.

If then the diverging rays or appendages of the thoracic and abdominal vertebræ of fishes (fig. 2, *a, a*), of reptiles (fig. 3, *a, a*), and of birds (fig. 4, *a, a*), be serial repetitions of the more developed appendage of the scapulo-coracoid arch, they must be 'rudimental limbs,' future possible or potential arms, legs, wings or feet.

To become recognised specially as a 'limb,' it needs only that the diverging appendage carry out beyond the surface a fold or sheath of the integument, which it may be able by its muscles to move and make react upon the ambient medium or supporting surface. Hence the simple filamentary ventral appendages of the lepidosiren and the blenny are acknowledged by Naturalists as hinder limbs. In Zoology, however, not more than two pairs of limbs are recognised in the Vertebrate series. But with our present knowledge of the nature of limbs, we are stimulated to inquire whether there be no other diverging appendages of hæmal arches similarly developed and deserving that name?

F

If we examine the segments of the skeleton in advance of the occipital one, which are represented by their inferior or hæmal arches in the lepidosiren *, we shall find that each of them developes its pair of appendages (23, 34, 37) : that (37) from the hyoid (parieto-hæmal) arch projects freely outwards as a simple osseous unsegmented ray : that (34) from the tympano-maxillary (fronto-hæmal) arch has a like form and condition. The diverging appendage of the maxillary (naso-hæmal arch) is anchylosed as a (pterygoid) process, and repeats the function of the costal appendages in birds by attaching the arch from which it diverges to the next in succession. The free moveable appendages of the tympanic and hyoid arches carry out with them a fold of integument ; but, like the ventral and pectoral fins of the lump-fish (*Cyclopterus*), they combine to support the same fold or rudimental fin, which is denominated the ' operculum ' ; but this is essentially a limb, or rather two cephalic limbs conjoined, which, in fact, are specially distinguished in ichthyology as the ' opercular ' and ' branchiostegal ' flaps.

An osseous fish, therefore, of the type of that the skeleton of which is represented in Pl. I. fig. 2, is an ' octopod,' and six of its limbs belong to the head. With regard to the third pair, which are more especially organized for locomotion, whatever be their ultimate development and destination, they always first appear in the simple form in which they are permanently arrested in the lepidosiren, viz. as an unbranched ray. Development simply augments its length and the number of its segments in the lepidosiren : in other fishes the jointed ray is multiplied : sometimes a segment is modified between the terminal rays and the supporting arch, which may be recognised as a carpus (*Lophius*†): sometimes a second seg-

* Lectures on the Comparative Anatomy and Physiology of the Vertebrate Animals, 8vo, 1846, p. 79, fig. 27, 19, 32, 40.

† *Ib.* p. 121, fig. 40, 56.

ment is interposed between the ' carpus' and the arch, which may be recognised as the 'antibrachium' (Pl. I. fig. 2, 54, 55) : but the development of definite segments in the longitudinal direction of the limb does not go beyond this point in fishes. Far otherwise, however, is it in respect of development in breadth ; multiplication of digital rays is the characteristic type of the pectoral extremity in the piscine class, and reaches its maximum in the species which have accordingly received the emphatic denomination of ' Ray-fishes' (*Raiidæ*). And this mode of complication of the pectoral member by vegetative repetition of like parts, is a striking and instructive instance of the special divergence of the piscine branch from the common Vertebrate stem.

In order to follow the development of the diverging appendage of the occipital hæmal arch through other ramifications of that stem, we must retrace our steps to the species in which it retains its embryonic state and represents its archetypal character. The first step in development from the primitive type of the pectoral ray of the lepidosiren is made by the *Amphiuma didactylum* (Pl. I. fig. 8). The hæmal arch is detached from the neural arch, and slightly displaced backwards ; but the pleurapophysis (*pl*, 51) retains its simple rib-like form and position, slightly inclining outwards below from the vertical line. The hæmapophyses (*h*, 52) do not pass beyond the state of gristle, but are much expanded : they resemble in their histological condition their homotypes, called ' cartilages of the ribs,' in the thorax of man. If the study of the essential nature of the detached inverted arch so formed had been begun at this point and compared with that of the Vertebrates lower in the scale, no doubt, I apprehend, would have been entertained as to the detachment of such hæmal arch in the amphiuma being a deviation from type, and its attachment to the rest of its segment in the lepidosiren and osseous fishes as being a retention of the typical structure : this condition would have been in point of fact the rule, and the other the exception.

So likewise with respect to the diverging appendages, *a a*, of the occipito-hæmal arch of the amphiuma : if the anatomist had observed them with a previous knowledge only of the lepidosiren and other fishes, the bones 54, 55 and 57 would doubtless have been regarded and described only as bifid segments of the primitive simple ray. But the parts having been originally studied from a higher point in the animal series, where the homologues of those segments by virtue of their special developments in adaptation to special functions had obtained special names, those names were naturally transferred to their simplified homologues in the appendage recognised as the anterior limb or extremity of the amphiuma : the proximal single segment 53 was described as ' humerus,' the ossified divisions of the next segment as ' radius ' and ' ulna ' (54 and 55), the terminal bifurcation (57) as the ' digits.' This extreme instance of the exemplification from ' special homology' of the unity of the plan upon which the limbs of the vertebrate animals have been constructed is a perfectly true one.

Cuvier has most accurately assigned their special names to each of the parts of the fore-limb of the amphiuma in his celebrated memoir*. All that I would ask of the most devoted disciple of the school of ' positive facts ' is to reciprocate ; to grant the inference as to the signification of the parts arrived at by their study in the ascending route of inquiry, which the homologist is ready to give to the determinations of the special character of the parts which have been obtained by comparisons pursued descensively from man : in other words, to admit that the whole (53–57)

* " Dans ces deux figures *a* est l'*omoplate*, *b* les plaques sternales cartilagineuses formées probablement des *os coracoïdiens*; *c* l'*humerus*, suivi du *cubitus* et du *radius* qui portent un *carpe* cartilagineux et deux os *metacarpiens* et *phalangiens* osseux." Mémoire lu à l'Académie des Sciences, le 13 Novembre 1826, p. 15.

in the amphiuma (fig. 7) may be the homologue of the ray
(53–57) in the lepidosiren (fig. 6) ; that this may answer to
the ray (53–57 *a*) in the fourth segment of the archetype
(fig. 1) ; and that such ray is repeated in the diverging appendages, *a a,* of the succeeding segments of the skeleton :
whereby he will be led to the recognition of the essential
nature of the limbs as developed diverging appendages of the
hæmal arches of vertebræ, and the fore-limbs as being such
appendages of the occipital vertebra*. The facts are not
less ' positive' than they were before, only they cease to be
empirical and become intelligible.

I need not trespass further on the time of this distinguished audience, by adding instances of the complication
and concomitant powers of these appendages to those that
have already been illustrated at the commencement of the
present discourse ; I will only repeat, that the adaptive developments of the radiated appendage of the occipital hæmal
arch reach their maximum in man, and the distal segment
of the appendage constitutes in him an organ which the
greatest of ancient philosophers has defined as the " fit
instrument of the rational soul ;" and which an eminent
modern physiologist has described " as belonging exclusively to man—as the part to which the whole frame must
conform †." And these expressions give no exaggerated

---

* The want of connection of a peripheral piece, at its peripheral end
or border, appears to be one condition of its greater extent of variety of
form and proportion than in the more central pieces of a natural segment. There is nothing to restrain its luxuriant development from a
simple spine to a plate, to a divided plate with intercalations, &c., or
to a lengthened segmented ray bifurcating into additional segments
with indefinite modifications of these. Always, however, it is to be remembered that the polarising forces which tend to shoot out particle
upon particle after the pattern of dendritic corals, lichens or crystals,
are so controlled by the antagonising principle of adaptation, that the
radiating growth is always checked at that stage and guided to that
form which is best suited to the needs and habits and sphere of life of
the species.

† Sir Charles Bell, Bridgewater Treatise, 1833, pp. 16, 18.

Μόνον δὲ καὶ ἀμφιδέξιον γίγνεται τῶν ἄλλων ζώων ἄνθρωπος.—*Aristotle.*

idea of the exquisite mechanism and adjustment of its parts.

It is no mere transcendental dream, but true knowledge and legitimate fruit of inductive research, that clear insight into the essential nature of the organ, which is acquired by tracing it step by step from the unbranched pectoral ray of the lepidosiren to the equally small and slender but bifid pectoral ray of the amphiuma, thence to the similar but trifid ray of the proteus, and through the progressively superadded structures and perfections in higher reptiles and in mammals. If the special homology of each part of the diverging appendage and its supporting arch are recognisable from Man to the fish, shall we close the mind's eye to the evidences of that higher law of archetypal conformity on which the very power of tracing the lower and more special correspondences depend?

Until the alleged facts (pp. 50, 53) are disproved, demonstrating change of position to be one of the modifications by which parts of a natural and recognisable endoskeletal segment are adapted to special offices, and until the conclusions (pp. 54, 58) deduced from those facts are shown to be fallacious, I must retain the conviction that, in their relation to the vertebrate archetype, the human hands and arms are parts of the head—diverging appendages of the costal or hæmal arch of the occipital segment of the skull*.

---

* As another example of the new light and interest which a knowledge of general homology gives to the facts of abnormal anatomy in the human species, I may cite the remarkable case described by Sir C. Bell (op. cit. p. 52), of the boy 'born without arms,'—'but who had clavicles and scapulæ.' Here development was arrested at the point at which it is normal in the *Anguis, Pseudopus,* and some other limbless and snake-like lizards. The usual predominating development of the scapular appendage has bred so prevalent an idea of the subordinate character of the supporting arch, that the existence of the arch minus the appendage, is adverted to not without a note of surprise in the above-cited and other excellent works. General homology, however, teaches that a vertebral arch is a more constant and important part than its appendages; and, that, being anterior in the order of develop-

Having thus arrived at a solution of the question mooted at the outset of the present Discourse, viz. the signification, ' bedeutung' or general homology of locomotive members, and of the pectoral pair in particular, little needs to be added in respect of the pelvic pair of limbs.

When a bone in the skeleton of a dog, a horse, a mole, and a platypus, is proved to be the same or answerable bone to the one called ' scapula' in man, and is called by the same name; and when the answerable bone can be traced through birds, reptiles and fishes, where it is similarly recognised and indicated,—determined, in short, to be the namesake or ' homologue' of the human scapula; it follows, that, whatever other or higher proposition respecting the nature of that bone and its relations to the fundamental pattern of the vertebrate skeleton can be demonstrated by the sum of its characters in all applies individually to every form of the bone : and whatever element of the typical segment of the skeleton it may be recognised to be in an instance where the typical characters are best retained, such conclusion equally applies to the instances in which it is most metamorphosed. In other words, when the general homology of a bone is determined in any one species, the same is proved of its special homologues in every other species. If the scapula of the lepidosiren is the rib of the occipital vertebra, every other scapula must be the same element.

ment, it may be expected, in cases where development is arrested, whether normally in accordance with the nature of the species, or abnormally as an individual defect, to be present when the diverging appendages are absent. Sir Charles Bell, well recognising the primary function of the modified occipital rib in relation to breathing, observes, in reference to the above-cited case, " We would do well to remember this double office of the scapula and its muscles, that, whilst it is the very foundation of the bones of the upper extremity, and never wanting in any animal that has the most remote resemblance to an arm, it is the centre and point d'appui of the muscles of respiration, and acts in that capacity where there are no extremities at all !"—P. 52.

So, likewise, when a bone in a given natural segment of the skeleton is demonstrated to be answerable to another bone in another natural segment of the same skeleton, whatever higher homological proposition may be demonstrated of the one must apply to the other and to all its serial homologues, *i. e.* to all the bones that have been shown to stand in the same relations to their respective segments. The comprehension of these propositions will be facilitated by tracing the bones that have the same kind of marking, from segment to segment, in the diagrams of the skeletons in Pl. I.

No anatomist, I presume, could doubt or would contravene so plain a proposition as that the frontal bone or bones in their part of the skull repeat the general relations of the parietal bone or bones in their part, and those of the supraoccipital bone in its part of the skull; or that the basioccipital in its segment repeated the essential characters of the basisphenoid in the next segment.

Whatever element, therefore, of the typical segment the supraoccipital is demonstrated to be, the same must be true, not only of the supraoccipital in all other animals, but of the parietals and frontals in the succeeding segments of the same animal. And whatever element the basioccipital is demonstrated to be in its segment, the basisphenoid must be that element in the succeeding segment. In other words, when parts are demonstrated to be ' homotypes' or serial homologues, a general homology proved of one applies to all*.

A due appreciation of these rigorous deductions will leave no difficulty in dealing with the general homology of the hinder or pelvic limbs.

Anatomists with one consent admit that the ilium is the

---

* The terms ' general,' ' special,' and ' serial homology,' are explained and exemplified in my " Lectures on the Comparative Anatomy and Physiology of the Vertebrata," p. 48, and ' On the Archetype of the Vertebrate Skeleton,' p. 7.

homotype of the scapula; in other words, is the answer-able bone in the series of segments of the same skeleton: few, if any, doubt, and none with reason, that the ischium is the homotype of the coracoid, and the pubis of the cla-vicle. The ilium must therefore be, like the scapula, a ' pleurapophysis,' and the ischium and pubis, like the cora-coid and clavicle, must be ' hæmapophyses' ; for whether the serial homology just enunciated in regard to the latter bones be accepted or inverted, and the pubis be viewed as the homotype of the clavicle or coracoid, matters not ; they are the same elements of the hæmal arch in the abs-tract; their general homology is, therefore, the same.

The serial homology of the pectoral and pelvic extremi-ties has been so fully discussed, that no more need be added on that head to prove their general homology. If the pectoral members have been demonstrated to be develop-ments of the diverging appendages of the hæmal arch, the pelvic members must necessarily be the same elements of the typical vertebra.

This conclusion as to the general homology of the hind-limbs would, however, have been arrived at if the study of the Nature of Limbs had been commenced with them, and had been illustrated by their morphological and develop-mental characters, independently of the light reflected from serial homology. But as it may serve to establish con-fidence in the truth of the evidence by which such genera-lizations are arrived at, I will briefly adduce some of the more striking facts or links in that independent chain of reasoning.

The most simple and elementary condition of the com-plete and normally connected pelvic arch and appendages is found in the lowest forms of crawling reptiles, viz. the perennibranchiate *Batrachia*. Here is a side view of the parts in the *Menopoma* (fig. 11), and fig. 10 in Pl. I. gives a direct front-view of them in the *Proteus*. In fig. 11 the three anterior vertebræ which answer in position to the

'lumbar' vertebræ in Pl. I. fig. 3, differ chiefly in having
pleurapophyses (P*l*) articulated and not anchylosed to the

Fig. 11.

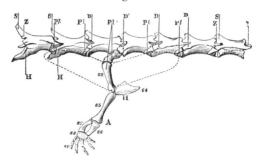

Pelvic vertebra and appendage with contiguous vertebræ (Menopome).

ends of the diapophyses (D) ; but the ribs are rudimental,
and the archetypal segment is departed from by defect of
the hæmal arch. In the next vertebra the diapophysis
(D') is thickened, and the pleurapophysis (P*l*) elongated
and divided into two parts like the occipital pleurapophysis
in the fish (Pl. I. fig. 2, 50, 51): and the second rib-like
piece (62, fig. 11) is joined by its lower end to a broad
partially ossified cartilage (64), the hæmapophysis, which
meets and joins its fellow, completing a hæmal arch and
restoring the vertebra in question to the typical character.
A radiated appendage, A, moreover, diverges on each side
from the articulation between 62 and 64, and forms the
hind-limb. Now the special homology of this limb with
the undivided filamentary appendage similarly situated in
the lepidosiren (Pl. I. fig. 9) and with the ventral fins of
fishes, in the descending series, and with the hind-limb of
other reptiles, of birds and of mammals in the ascending
series, is unmistakeable, and, I believe, is generally admitted : so that comparative anatomists have not hesitated
to call the rib-like bone (62) 'ilium,' and the cartilage (64)
'ischium' or 'pubis' in the menopome.

The correspondence of the segment of the endoskeleton in the menopome D', P*l*', H, A, with the typical vertebra, as illustrated by fig. 7, p. 42, is such, that any other explanation of its essential nature than as a representative or repetition of such fully-developed segment or vertebra, would be plainly contrary to nature. The chief modification has its seat in the most peripheral part, viz. the appendage A, as compared with its simple homologue in the thoracic segment of the bird (fig. 7, *a*). If 62 and 64, fig. 11, are to be regarded as strangers to the vertebral system, new parts introduced for special purposes, and not as normal elements modified for special purposes, I am at a loss to know on what principles, or by what series of comparisons we can ever hope to attain to the higher generalizations of anatomy, or discover the pattern after which the vertebrate forms have been constructed.

It may be said that the arch which they constitute performs a new function, inasmuch as it sustains a locomotive limb which reacts upon the ground. But this new function arises in the menopome, rather out of the modifications of the appendage than of the arch itself. In so far as the mere support of the appendage is concerned, the inverted or hæmal arch (P*l*', H) performs no new function, but one which is common to such arches in the thorax of birds, and to the less completely ossified homologous arches in the abdomen of fishes, where moreover the simple diverging appendages (Pl. I. fig. 2, *a*, *a*) do give attachment to the muscles of locomotion.

Comparing, then, the hæmal arch in question with that of the typical vertebra (fig. 7, p. 42), we find that, like the scapulo-coracoid arch in fishes (Pl. I. fig. 2, and fig. 7, 50, 51, 52), its parts are open to two interpretations. The upper piece of P*l*' may be the whole pleurapophysis, the lower (62) the hæmapophysis, and the part, 64, the half of an expanded and bifid hæmal spine: or P*l*' with 62 may be two portions of a teleologically compound pleurapophysis, and 64 the hæmapophysis, which would join with its fellow without, or with a mere rudiment of, a hæmal spine intervening. From the analogy of the scapulo-coracoid arch in fishes, which is proved by its modifications in higher animals to want the hæmal spine, we might infer that such was the condition and true interpretation of the correspondingly simple pelvic arch under consideration, and the inference is confirmed by the undivided condition of the pleurapophysis or 'ilium' in the proteus (Pl. I. fig. 10, 62), as well as by the position from which the appendage or limb diverges. But the general relation of this arch to the hæmal one of the typical segment is not affected by the alternative.

In ascending from the proteus and menopome to the crocodile (Pl. I. fig. 3), we find the homologue of 62 broader than it is long, and articulated to the thickened proximal portions of the pleurapophyses of two vertebræ; and we observe, likewise, the pelvic arch completed below by two pairs of hæmapophyses: for the anterior pair the name of 'ossa pubis' is retained; for the posterior pair that of 'ischia.' In general homology these bones complete, as hæmapophyses, the two vertebral segments modified to form the sacrum of the crocodile; and the intermediate connecting piece (ilium) might be interpreted, as either the confluent distal portions of the pleurapophyses of both vertebræ, or as an expansion of one such portion, answering to 62 in the menopome, and intruding itself between the stunted pleurapophysis and distant hæmapophysis of the other sacral vertebra in the crocodile.

The doubt thus left in the prosecution of the inquiry by the successive steps of special homology is resolved by the light of serial homology. The ilium is the homotype of the scapula; it parallels the scapula in the nature and extent of its morphological modifications; and, since the scapula under its extremest expansion is proved to be the development of one and the same element, viz. the lower portion of the divided occipital pleurapophysis in the fish, so likewise must the broadest ilium be considered to be the development of the same portion of one pelvic pleurapophysis. Its broad expanded modification is its common morphological character in Mammalia; and its position becomes less vertical than in reptiles, and more oblique; and in both these particulars it resembles the scapula. In Plate I. 62 shows its common form in quadrupeds, fig. 5, and 62 in fig. 6 its still broader proportions in man, in relation to the extent of surface required by the muscles that sustain the upright trunk upon the diverging appendage of the arch.

The second sacral vertebra in man is complete; but its pleurapophysis is in two pieces, as in the menopome: the proximal piece coalesces with the neural arch and forms the so-called 'transverse process' of the vertebra; the distal or lower portion is expanded to form the so-called 'ilium' (62). The hæmapophysis (63) coalesces with that of the preceding vertebra (64), and with its own pleurapophysis (62).

The first sacral vertebra has its hæmapophysis (64, called 'pubis') ossified, but separated from its proper pleurapophysis by the expanded (iliac) portion of that of the succeeding vertebra (62), with which it coalesces, as well as with the succeeding hæmapophysis (63, 'ischium'). The short and thick pleurapophyses of the third sacral vertebra also articulate, in the adult, with the expanded distal portion of those of the second sacral vertebra; but these (iliac bones) are restricted in infancy and early childhood to their connections with the first and second sacral vertebræ, which connections are permanent in most reptiles.

In the bird the modification of the vertebral segments at the posterior region of the trunk in relation to the transference of the whole weight of the body and fore-limbs (wings) upon the hind-limbs, is greater and more extensive than in man, and the essential nature of the pelvic arch is still more masked. By the extreme expansion of the element 62, fig. 4, Pl. I., it is brought into connection with the homologous stunted or proximal ends of pleurapophyses of several contiguous segments, besides its own proximal piece, in the manner indicated by the dotted line in the menopome (fig. 11, p. 74).

Now, if the ilium, so expanded, were interpreted as the coalesced complementary portions of all the short pleurapophyses with which it articulates, its nature would be very similar to that which Oken has attributed to the scapula. But its ossification radiates, as in the simple rib-like ilium of the menopome, from a common centre: there are no corresponding multiplications of hæmapophyses below; these are restricted in the pelvis of all animals to the number which they present in the crocodile. And since the scapula has been proved to be, under its most expanded form, the homologue of a single pleurapophysis, so also its homotype, the ilium, must be regarded as maintaining, under every variety of form and proportion, the same fundamental singleness of character which it presents on its first appearance in the perennibranchiate batrachian.

The rudimental hind-limb in serpents is suspended in the flesh and attached only indirectly by one of its simple muscles to the bifurcate and shortened rib of the anterior caudal vertebra: that pleurapophysis, therefore, is but little more modified than the one which represents the ilium in the proteus, and the diverging appendage is as simple as in the lepidosiren, consisting of fewer joints, even where best developed as in the boa, and being reduced to a single osseous style in the slow-worm.

In the air-breathing Vertebrata the typical character of the pelvic arch is progressively disguised by excess of development as we ascend from the low point at which we commenced its analysis; and, in descending below the same low point to the water-breathing class, we find the pelvic arch deviating from its typical character by defective development. And for this indeed we might have been prepared by the consideration of the close relation which this arch and its appendages bear to locomotion and support of the body on dry land.

In fishes (Pl. I. fig. 2) we find the fin-like homologue of the hind-limbs (v) radiating from a bone (63) which converges to its fellow at the median line, and which is recognised by some anatomists as the 'pubis'; by others, and with better reason, as the 'ischium'; but which is evidently the same abstract vertebral element, viz. a 'hæmapophysis.' The ilium, if it be developed, retains its character as a short, free, or false rib, like 62, its development not proceeding so far as to effect the normal junction with the hæmapophysis. This lower element therefore of the arch being liberated from its typical connections, has no fixed position in the class of Fishes, but in some existing species and all the primæval forms of fishes, keeps near its proper segment, as at v, fig. 2. Pl. I.; and that this is its true position in relation to the archetype, is significantly indicated by the fact that all the fishes in the geological formations anterior to the chalk are 'abdominal.' In certain species of the actual creation the ventral fins advance to the place marked v' in fig. 2, the ischium elongating to join the coracoid, just as one detached costal cartilage is suspended from another at the back of the thorax in certain birds. By the shortening of the attached ischium, the ventral fins, in other existing fishes, are brought forwards to v''.

In Pl. I. fig. 9, the elementary condition of the hind-limbs in the vertebrata is shown in nature in a back view of the pelvic vertebra of the *Protopterus* or lepidosiren. The letters signify the general and the figures the special homologies of the parts. The apical elements (hæmapophyses or ischia, 63) of the inverted arch are detached from the basal ones (pleurapophyses or 'ilia,' 62) and from the rest of their segment, and carry with them the diverging appendages (65–69), as in all other fishes.

The true meaning and nature of this piscine condition

is beautifully illustrated if we now return, by one step in advance, to the point from which we commenced the analysis of the pelvic arch and limbs.

In fig. 10. Pl. I. *e. g.* the hæmal arch retains its natural connections with the rest of its vertebra, and henceforth preserves them, with a few exceptions (*Enaliosauria* and *Cetacea*), in all the air-breathing classes, up to and including man.

In respect of the modification by displacement, the numerical examples of adhesion to and of departure from type are reversed in the pelvic segment, as compared with the occipital one. Mammals, birds and reptiles show the rule of connection, and fishes the exception, typically as well as numerically. There has been, therefore, no difficulty or discrepancy of opinion in regard to the homology of the detached hæmal arch and its appendages in fishes. Cuvier saw in 63, fig. 2, the representative of the 'os innominatum' or 'os du bassin;' and, notwithstanding the degree of displacement to which such rudiment of a pelvis, with its pelvic members, were subject in fishes, Linnæus had as little hesitation in recognizing in the ventral fins the homologues of hind-limbs wherever they were placed. When in their normal position, as at **v**, fig. 2, they characterized the 'abdominal' fishes; when advanced to beneath the pectoral fins, as at **v'**, they characterized the 'thoracic' fishes; when still more advanced, as at **v''**, they characterized the 'jugular' fishes. The species in which the ventral fins were absent were 'apodal,' in the philosophic language of the immortal Swede. He knew them to be hind-feet under their webbed disguise.

Now all that is here required, in regard to the determination of the locomotive members, is, that no more value be given to the character of detachment and change of place in regard to the scapular arch and its appendages than Linnæus allowed in the case of the pelvic arch and its appendages.

The arms are shifted to and fro in the bodies of the air-breathing vertebrates, the legs in those of the water-breathing vertebrates : the arch supporting the arms is fixed in its true place in fishes, and the arch supporting the legs retains its true place in the higher classes ; only it is often necessary that it should be so developed as to be applied to many segments besides the one to which it properly belongs. In the proteus (fig. 10), however, the ilium (62) retains its simple primitive rib-like form, just as the scapula (51) does in fig. 8 ; and it is connected, as we saw likewise in the menopome (p. 74, fig. 11), to its own vertebra exclusively.

Wherever either arch with its appendages may be situated, it is in its best possible place in relation to the exigences and sphere of life of the species. It is only when we consider its relations to the ideal exemplar that we are compelled to use the terms expressive of rule and deviation.

When a truth is arrived at, the conception of it is clear, and it can be expressed plainly and intelligibly. When the inquirer stops short of, or goes astray from, his aim, yet thinks he has gained it, he can only have a distant and cloudy notion of it, and his definitions must partake of the obscurity of his conceptions. When Oken, after recognising in the scapula an aggregate of cervical ribs, afterwards defined the limbs as ' liberated ribs,' we perceive that neither his notion of ribs, or of the scapular arch, or of limbs, could be clear. When Carus explains the nature of limbs by calling them ' tertiary vertebræ,' we discern the same obscurity in his idea of the primary segments of the skeleton.

According to my definition of a vertebra, we recognise in it the centrum, the neural arch, the hæmal arch, and the appendages diverging or radiating from the hæmal arch. The centrum, though the basis, is not less a part of a vertebra than are the neurapophyses, hæmapophyses, pleur-

G

apophyses, &c.; and each of these parts is a different part from the other: to call all these parts 'vertebræ' is in effect to deny their differential and subordinate characters, and to voluntarily abdicate the power of appreciating and expressing them. The terms 'secondary' or 'tertiary vertebræ' cannot, therefore, be correctly applied to the parts or appendages of that natural segment of the endoskeleton to the whole of which segment the term 'vertebra' ought to be restricted.

So likewise the term 'rib' may be given to each moiety of the hæmal arch of a vertebra; although I would confine such special appellation to the pleurapophyses when they present that long and slender form characteristic of the thoracic abdominal region, viz. that part of such modified hæmal or costal arch to which the term 'vertebral rib' is applied in anatomy and the term 'pars ossea costæ' in anthropotomy: but, admitting the wider application of the term 'rib' to the whole hæmal arch under every modification, yet the bony diverging and backward projecting appendage of such rib or arch is something different from the part supporting it.

ARMS AND LEGS, therefore, ARE DEVELOPMENTS OF COSTAL APPENDAGES. They are not ribs that have become free: although liberated ribs may perform analogous functions, as in the serpents and the *Draco volans*.

Having thus attained the end proposed as the subject of the present discourse, let us finally apply the few minutes that remain of the allotted time to the contemplation of the abstract Archetype skeleton, as illustrated in fig. 1, Pl. I.

The pectoral and ventral limbs are there exhibited as the appendages of the fourth and twenty-sixth segments: the occipital segment is always the fourth, the pelvic segment has no fixed place or numerical symbol. The lepidosiren realizes, very closely, the primitive condition of the limbs, as exemplified in the ideal archetype.

We have been accustomed to regard the vertebrate animals as being characterized by the limitation of their limbs to two pairs, and it is true that no more diverging appendages are developed for station, locomotion and manipulation. But the rudiments of many more pairs are present in many species. And though they may never be developed as such in this planet, it is quite conceivable that certain of them may be so developed, if the Vertebrate type should be that on which any of the inhabitants of other planets of our system are organized.

The conceivable modifications of the vertebrate archetype are very far from being exhausted by any of the forms that now inhabit the earth, or that are known to have existed here at any period.

The naturalist and anatomist, in digesting the knowledge which the astronomer has been able to furnish regarding the planets and the mechanism of the satellites for illuminating the night-season of the distant orbs that revolve round our common sun, can hardly avoid speculating on the organic mechanism that may exist to profit by such sources of light, and which must exist, if the only conceivable purpose of those beneficent arrangements is to be fulfilled. But the laws of light, as of gravitation, being the same in Jupiter as here, the eyes of such creatures as may disport in the soft reflected beams of its moons will probably be organized on the same dioptric principles as those of the animals of a like grade of organization on this earth. And the inference as to the possibility of the vertebrate type being the basis of the organization of some of the inhabitants of other planets will not appear so hazardous, when it is remembered that the orbits or protective cavities of the eyes of the Vertebrata of this planet are constructed of modified vertebræ. Our thoughts are free to soar as far as any legitimate analogy may seem to guide them rightly in the boundless ocean of unknown truth. And if censure be merited for here indulging, even for a moment, in pure

G 2

speculation, it may, perhaps, be disarmed by the reflection
that the discovery of the vertebrate archetype could not
fail to suggest to the Anatomist many possible modifications of it beyond those that we know to have been realized
in this little orb of ours.

The inspired Writer, the Poet and the Artist alone have
been privileged to depict such.

Something also I would fain add with a view to remove
or allay the scruples of those who may feel offended at any
expressions that seem to imply that any part or particle of
a created being could be made in vain.

Those physiologists who admit no other principle to
have governed the construction of living beings than the
exclusive and absolute adaptation of every part to its function, are apt to object to such remarks as have been offered
regarding the composition of the skeleton of the whale's
fin and of the chick's head, that 'nothing is made in vain;'
and they deem that adage a sufficient refutation of the
idea that so many apparently superfluous bones and joints
should exist in their particular order and collocation in
subordination to another principle; conceiving, quite gratuitously in my opinion, the idea of conformity to type to
be opposed to the idea of design.

But let us consider the meaning which in such discussions is commonly attached to the phrase 'made in
vain.' Were the teleologist to analyse his belief in the
principle governing organization, he would, perhaps, find
it to mean, that so far as he can conceive of mechanism
directly adapted to a special end, he deems every organic
mechanism to have been so conceived and adapted. In
a majority of instances he finds the adaptation of the
organ to its function square with his notions of the perfection of a machine constructed for such an end; and
in the exceptional cases, where the relation of the
ascertained structure of an organ is not so to be understood, he is disposed to believe that that structure may

be, nevertheless, as directly needed to perform the function, although he perceives that function to be a simple mechanical action, and might conceive a more simple mechanism for performing it. The fallacy perhaps lies in judging of created organs by the analogy of made machines; but it is certain that in the instances where that analogy fails to explain the structure of an organ, such structure does not exist 'in vain' if its truer comprehension lead rational and responsible beings to a better conception of their own origin and Creator*. Our philosophic poet felt that—

> " 'Tis the sublime of Man,
> Our noontide majesty, to know ourselves
> Parts and proportions of a wondrous whole."—*Coleridge.*

Nor could the ignorance of this truth be without its benumbing and bewildering influence on the human mind.

The learned Cudworth tells us that—" The Democritick Atheists reason thus: If the World were made by any Antecedent Mind or Understanding, that is, by a Deity; then there must needs be an 'Idæa' and 'Exemplar' of the whole world before it was made, and consequently actual knowledge, both in order of Time and Nature, before Things."

But these Democritans arguing of knowledge as it is got by our finite minds, and ignorant of any evidence of an ideal Archetype for the world or any part of it, concluded that there could be no knowledge or mind before the world, as its cause. And in the same spirit Lucretius asks:

> " Exemplum porro gignundis rebus et ipsa
> Notities hominum Divis unde insita primùm,
> Quid vellent facere ut scirent animoque viderent?"

Now, however, the recognition of an ideal Exemplar for the Vertebrated animals proves that the knowledge of such

---

* " L'ensemble de tous les ordres de perfections relatives, compose la perfection absolue de ce tout. L'Unité du dessein nous conduit à l'unité de l'intelligence qui l'a conçu."—*Bonnet, Contemplations de la Nature.*

a being as Man must have existed before Man appeared. For the Divine mind which planned the Archetype also foreknew all its modifications.

The Archetypal idea was manifested in the flesh, under divers such modifications, upon this planet, long prior to the existence of those animal species that actually exemplify it.

To what natural laws or secondary causes the orderly succession and progression of such organic phænomena may have been committed we as yet are ignorant. But if, without derogation of the Divine power, we may conceive the existence of such ministers, and personify them by the term 'Nature,' we learn from the past history of our globe that she has advanced with slow and stately steps, guided by the archetypal light, amidst the wreck of worlds, from the first embodiment of the Vertebrate idea under its old Ichthyic vestment, until it became arrayed in the glorious garb of the Human form.

DESCRIPTION OF PLATE I.*

THIS Plate includes diagrams of the ideal pattern or archetype of the vertebrate endoskeleton, and of the modifications of it characteristic of the four great divisions of the vertebrate subkingdom, viz. fishes, reptiles, birds, and mammals, and also of man.

In each figure the parts or 'elements' of the four anterior segments—the seat of the chief modifications in relation to the lodgement of the brain, the action of the jaws and tongue, and the interposition of the sense-organs—are numbered as in the column of *Nomina* in the Plate.

As the four anterior segments of the neural axis are called collectively 'brain' (*encephalon*), so the four corresponding segments of the vertebral axis are called collectively 'skull' (*cranium*). The head therefore is not otherwise a repetition of the rest of the body, than insofar as each segment of the skull is a repetition or 'homotype' of every other segment of the body ; each being subject to modifications which give it its individual character, without obliterating its typical features. So neither are the 'arms' and 'legs' repeated in the head in any other sense than as the cranial segments may retain their diverging appendages. The 'forelimbs' are actually such appendages of the occipital vertebra, which undergo modifications closely analogous to those of the appendages of the pelvic segment or 'hind-limbs.' And inasmuch as in one class the pelvic appendages, with their supporting hæmal arch, are detached from the rest of their segments, and subject to changes of position (fig. 2, V, V', V''), so also in other classes the appendages of the occipital segment are liable to be detached with their sustaining hæmal arch, and to be transported to various distances from their proper centrum and neural arch.

* The Plate and the description are taken from my work ' On the Archetype and Homologies of the Vertebrate Skeleton,' 8vo, Van Voorst, 1848.

The head therefore is not a virtual equivalent of the trunk, but is only a portion, *i. e.* certain modified segments, of the whole body.

The jaws are the modified hæmal arches of the first two segments; they are not 'limbs' of the head*.

The different elements of the primary segments are distinguished by peculiar markings :—

the neurapophyses by diagonal lines, thus :—

the diapophyses by vertical lines :—

the parapophyses by horizontal lines :—

the *centrum* by decussating horizontal and vertical lines :—

the pleurapophyses by diagonal lines :—

the hæmapophyses by dots :—

the appendages by interrupted lines :—

the neural spines and hæmal spines are left blank.

In certain segments the elements are also specified by the initials of their names, as in the third segment in fig. 1, and the fourth in fig. 2, for example :—

> *ns* is the neural spine.
> *n* is the neurapophysis.
> *pl* is the pleurapophysis.
> *c* is the centrum.
> *h* is the hæmapophysis.
> *hs* is the hæmal spine.
> *a* is the appendage.

Fig. 1. Ideal pattern or archetype of the vertebrate endoskeleton, as shown in a side-view of the series of typical segments, osteocommata, or 'vertebræ' of which it is composed, the ex-

---

* These are counter-propositions to Oken's "Der Kopf ist der ganze Rumpf mit allen seinen Systemen," &c., 'Lehrbuch der Natur-philosophie,' 8vo, p. 300, 1843. See the Translation of this Work published by the Ray Society. "The head is the whole trunk with all its systems, the brain is the spinal marrow, the skull the vertebral column, the mouth intestine and abdomen, the nose lung and thorax, the jaws are members or limbs."—P. 364, Prop. 2072.

treme ones indicating the commencement of those modifications, which, according to their kind and extent, impress class-characters upon the type.

The four anterior neurapophyses, 14, 10, 6, 2, give issue to the nerves, the terminal modifications of which constitute the organs of special sense.

The first or foremost of these is the organ of smell (18, 19), always situated immediately in advance of its proper segment, which becomes variously and extensively modified to inclose and protect it.

The second is the organ of sight (17), lodged in a cavity or 'orbit' between its own and the nasal segment, but here drawn above that interspace.

The third is the organ of taste, the nerve of which (gustatory portion of the trigeminal) perforates the neurapophysis (6) of its proper segment (*vertebra parietalis* seu *gustatoria*), or passes by a notch between this and the frontal vertebra, to expand in the organ which is always lodged below in the cavity called 'mouth,' and is supported by the hæmal spine (41, 42) of its own vertebra.

The fourth is the organ of hearing (16), indicated above the interspace between the neurapophysis of its own (occipital) and that of the antecedent (parietal) vertebra, in which interspace it is always lodged ; the surrounding vertebral elements being modified to form the cavity for its reception, called 'otocrane.'

The mouth opens at the interspace between the hæmal arches of the anterior and second segment ; the position of the vent varies (in fishes), but always opens behind the pelvic arch (Pv) when this is ossified.

Outlines of the chief developments of the dermoskeleton, in different vertebrates, which are usually more or less ossified, are added to the endoskeletal archetype ; as, *e. g.* the median horn supported by the nasal spine (15) in the rhinoceros ; the pair of lateral horns developed from the frontal spine (11) in most rumi-nants ; the median folds (DI, DII) above the neural spines, one or more in number, constituting the 'dorsal' fin or fins in fishes and cetaceans, and the dorsal hump or humps in the buffaloes and camels ; similar folds are sometimes developed at the end of the tail, forming a 'caudal' fin, C, and beneath the hæmal spines, constituting the 'anal' fin or fins, A, of fishes, or the subcaudal dermo-adipose tumour of the Cape-sheep.

Fig. 2. Typical skeleton of a fish (class *Pisces*). The plane of the anterior hæmal arch (20, 21, 22) is here raised to parallel with the axis of the trunk, and its apex or spine (22) is modified and developed so as to articulate with the neural spine (15) of the same segment, which thus becomes closed anteriorly; both 22 and the hæmapophysis 21 are developed downwards and backwards in relation to the protractile and retractile motions of the arch; and, for the purpose of associating these motions with corresponding ones of the succeeding hæmal arch, the diverging appendage is subdivided (23 and 24) and developed so as to articulate with the pleurapophysis (28) of the next arch; a rudiment of an appendage (26) is attached in some fishes to the hæmapophysis (21) of the nasal segment, but it will be observed that no new element is added to the hæmal arch; and, although the *Lepidosteus* offers an exceptional instance of subdivision of the pleurapophysis (21), that kind of modification is usually restricted to the diverging appendage.

In the next segment the hæmal arch has been the seat of unusual growth, but retains more of its normal position and attachments. Its weight and that of the appendages it supports have required an extension of the proximal articulation of its pleurapophysis (28 *a*) from its proper parapophysis (12) backwards to the next parapophysis (8); and the pleurapophysis itself is subdivided into two, three, or four overlapping pieces for the final purpose explained in my Work "On the Archetype," &c. p. 112; but it is evident that no new element has been introduced, because the extremities of the subdivided pleurapophysis (28 *a* and 28 *d*) retain their normal connections, the one with the parapophysis (12), the other with the hæmapophysis (29, 30). This element is also subdivided, for the same final purpose as the pleurapophysis; and its squamous union with the hæmal spine (32) is retained. Yet the connections of 29 with the condyle of the pleurapophysis and of 32 with its fellow, forming the free apex of the inverted arch of the second segment, show that the complexity is the result of adaptive subdivision, and that no new part has been added to the typical elements as exhibited in the archetype (fig. 1, 29–32); every anatomist has recognised the bones so numbered in the fish as the homologue of the single (undivided or anchylosed) bone forming the lower jaw (29–32) of the mammal (fig. 5) and of man (fig. 6). In addition, therefore, to change of shape and proportion, the parts of the archetype may be modified by division and subdivision. And in this respect the pleurapophyses (28) and hæmapophyses (29, 30, 31) of the fish deviate further from the archetype than the same parts do in the warm-blooded vertebrates. Herein is manifested the early divergence to a special form for the lowest class, which the higher classes do not assume in passing towards their own types. The diverging appendages are the seat of such excess of subdivision with special development of the divided parts, as best to countenance the idea of a superaddition of new parts to the typical element; yet the most essential character of the diverging appendage is retained under its extremest modification, as where it forms the wing of the bird or the arm of man; viz. its connection by one extremity to a hæmal arch, and the free projection of the opposite subdivided extremity, carrying out with it a fold of integument. With regard to the diverging appendage of the hæmal arch of the second segment, its modifications are arrested at different stages of departure from the simple archetypal form (34–37, fig. 1), as explained at pp. 66 and 112 of my Work "On the Archetype," &c. The most common modification in bony fishes is that shown in fig. 2, where it is divided into two segments, and the second segment into three pieces (35, 36, 37), usually broad and flat, for the office explained at p. 112 of the same Work.

The parietal segment, or third counting backwards, has the hæmal arch (38–41) detached from its proper supporting parapophysis (s) by the backward development of 28 *a* of the preceding segment. This is the first example of another modification, viz. that of dislocation, sometimes accompanied by great change of place, which has tended most to obscure the essential nature of parts, and their true relations to the archetype. The principle of subdivision still manifests itself in the elements of the hæmal arch, especially in its spine, 41–43; and in a greater degree by a vegetative repetition of the 'appendage' (44), without departure from its primitive ray-like form.

The pleurapophysis of the occipital segment (50, 51) is divided into two, and its proximal end is usually bifurcate in fishes, articulating like the normal ribs of higher animals, by a 'head' and a 'tubercle' to two points of the neural arch of its segment.

Almost every stage of development and departure from the primitive type is manifested by the diverging appendage (54–57) up to the extent of modification attained by the typical osseous fish. The proximal segment is divided into two pieces (54 and 55), the next segment into four or more (56), and the last segment into a greater but variable number of pieces, retaining the elementary form of rays.

The Lepidosiren (fig. 7) is eminently instructive by the retention in the occipital vertebra of the primitive condition of the appendage, as shown in the archetype (fig. 1, 53–57) modified only by segmentation of the ray. The pleurapophysis of the arch (51) likewise retains its simple cylindrical form, and is articulated to its centrum, like the other ribs of the Lepidosiren, by an undivided head.

The hæmal arch of the fifth segment (first of the trunk) is commonly detached from its centrum and neural arch in fishes without being displaced backwards. The pleurapophysis (pl) is short and simple, sometimes expanded; the hæmapophysis (58, h) is simple, long and slender. When this arch supports an appendage it is a simple diverging ray.

All the succeeding abdominal segments of the fish have their hæmal arches incomplete by bone; the hæmapophyses and spines retaining the primitive fibrous condition. The pleurapophyses of most support diverging appendages in the form of simple undivided bony rays.

A part of the hæmal arch of a post-abdominal (pelvic) segment is ossified (63), and supports a more complex appendage (69) in the form of one, two or more jointed rays, which project beyond the surface and are enveloped by a fold of skin forming the 'ventral' fin, V, making a pair with the one on the opposite side. This partially ossified hæmapophysis articulates with its fellow by its anterior apex, forming a 'symphysis ischii' seu 'pubis'; and, in some fishes called 'abdominal,' it is connected to its proper pleurapophysis (62) by an aponeurosis representing its unossified continuation.

The remarkable degree to which one and the same part may be subject to the modification of change of position, is strikingly exemplified in this lower portion of the pelvic arch with its appendages in fishes. It may be moved forwards, so that the symphysis of the pelvic arch is brought into connection with that of the scapular arch; when, according to the length of the ossified parts of the pelvic hæmapophyses, the species is either 'thoracic,' as when the ventral fins are at V', or 'jugular,' when they are advanced to V''. The universally acknowledged and long recognised special homology of the hæmal arch and appendages of the pelvic vertebra, as the ' ventral fins' of fishes, under these changes of position, prepare us for the recognition of an analogous modification of the hæmal arch and appendages of the occipital vertebra in the higher classes of Vertebrata.

Beyond the abdomen the osseous and aponeurotic parts of the hæmal arches rapidly contract ; the progressively elongated parapophyses usually bend down and complete the inverted arch by their apical coalescence ; sometimes distinct pleurapophyses continue to form these arches ; sometimes these elements may be traced, anchylosed with their fellows of the opposite side, and with the coalesced extremities of the parapophyses. The bodies of a certain number of the terminal segments coalesce together in the typical osseous fishes, and support several neural and hæmal arches and spines, usually more or less expanded, and forming the basis of the caudal fin, C.

The ossified parts of the dermal median and symmetrical folds, constituting the dorsal (DI, DII), the anal (A), and caudal (C) fins, are added to the endoskeleton in fig. 2 ; in are the interneural spines ; dn the dermoneural spines ; ih the interhæmal spines ; dh the dermohæmal spines ; these form no part of the true vertebral skeleton, and are peculiar to fishes. The diagram of the modified cranial segments is not complicated by the outlines of the sense-capsules or mucodermal bones.

Thus, compared with the archetypal figure, the endoskeleton of the fish deviates by excess of development, manifested chiefly in the diverging appendages of the four anterior or cranial segments, and by arrest of development in most of the other segments ; but the principle of polaric or vegetative repetition greatly prevails, and more of the segments resemble one another than in any of the higher classes.

Fig. 3. The Crocodile is here taken as the type of the class *Reptilia*.

The hæmal arch of the anterior segment is now firmly fixed by excessive development, chiefly of its hæmapophyses (21), which have extended their attachments to all the elongated elements (13, 14 and 15) of their own neural arch. The diverging appendage (24) from the pleurapophysis (20) fixes the arch extensively to the centrums of the second and third segments : the appendage from the hæmapophysis (21) bifurcates ; one branch, divided into two pieces (26 and 27), connects the arch with the pleurapophysis (28) of the next segment; the other branch (25) extends the attachment to the parapophysis (12) of the same segment, and also to the appendage (24) of its own arch.

The pleurapophysis (28) of the frontal segment is undivided; it is represented as displaced and depressed; but in nature it still retains a small part of its connection with its proper pleurapophysis (12), although it is developed backwards so as chiefly to articulate with that (8) of the following segment: it supports no diverging appendage. The hæmapophysis (29–31) is more subdivided than in fishes, in relation to functions explained in my Work " On the Archetype," &c., pp. 122, 123.

The excess of development of the hæmal arch of the frontal vertebra is compensated by the defect of development of that of the parietal one (40, 41); and this constitutes the next great additional step in the deviation from the archetype. Only the hæmapophyses (40) are ossified: the hæmal spine, though much expanded and flattened, remains cartilaginous, and the pleurapophysis is represented by a feeble ligament. The whole arch is detached and displaced backwards, and its diverging appendages cease to be developed.

The tendency to retrogradation manifested by the preceding hæmal arches is carried out to a striking extent in that (51, 52) of the occipital segment (the fourth counting backwards): it overlaps the homotypal arches of the 8th to the 11th segments of the trunk: the ossified portions of both its constituent elements, 51 and 52, are simple; the hæmal spine, 52', is prolonged backwards. The diverging appendage manifests, in comparison with that in the fish, an additional segment (53), which is single; the segment of two pieces (54 and 55) is now the second. The rays of the distal segment are reduced to five in number, which is never afterwards exceeded in the vertebrate subkingdom. The dislocation and retrogradation of the posterior hæmal segment of the skull form the second chief additional feature of departure from the archetype, as compared with the skeleton of fishes. The third well-marked modification is the development of an inferior (cortical) portion of the body of the atlas (ca, x), distinct from the main part of that centrum (ca), which coalesces with that of the axis, and is commonly called its 'odontoid' process.

The nine segments that succeed the head resemble those of

fishes in the non-ossification of the hæmapophyses and hæmal spines, but deviate further from the archetype by the minor development of the pleurapophyses. These progressively elongate to the 12th vertebra, where the hæmal arch is completed by a hæmapophysis and hæmal spines.

The hæmapophyses are not so completely ossified as the pleurapophyses, and they are divided from these by the interposition of cartilaginous pieces, *a, a*; these pieces may be regarded either as dismemberments of the hæmapophyses, or as unossified parts of the pleurapophyses. The diverging appendages (*a, a*) are usually cartilaginous.

Beyond the 21st segment of the trunk* the pleurapophyses usually cease to be represented either by bone or cartilage: but the partially ossified hæmapophyses are continued to those of the pelvic segments, 64 and 63, *h*. In these segments the pleurapophyses reappear, and are divided into two parts, like those in the thorax: the proximal portions (*pl, pl*) are short and thick; the distal portions have either coalesced into one broad and thick plate (62 *pl*), or the distal portion of one pleurapophysis is still more remarkably developed and takes the place of two: this question is discussed at pp. 75 and 76. The two hæmapophyses (63, 64) are distinct and well-ossified. The diverging appendage (65–69) has been subject to the same kind and amount of development as that of the scapular arch (53–57). The first steps in the

---

* According to Cuvier, the pleurapophyses cease to appear after the 20th trunk-vertebra in *Crocodilus biporcatus*, and after the 19th in *Alligator lucius*. I allude to these differences for the purpose of remarking that the conformity of organization is greater than would appear at first sight from the formulæ of the vertebræ of the different species of crocodile cited in the Table at p. 220, tome i. Leçons d'Anatomie Comparée, 1835. The number of vertebræ from the atlas inclusive to the sacrum is the same in each species, as will be seen by the following extract:—

	Cervical.	Dorsal.	Lumbar.		
Crocodile à deux arêtes	7	13	4	=	24
Crocodile du Gange	7	14	3	=	24
Caiman à mus. de brochet	7	12	5	=	24

The difference in the dorsal and lumbar series depends merely on the ossification or otherwise of the pleurapophysial tendons or fibrous bases attached to the diapophyses of the 20th, 21st and 22nd vertebræ.

A slight change in the form and size of the pleurapophysis is all that distinguishes the first dorsal from the last cervical vertebra in the Cuvierian Table.

progression of this metamorphosis from the primitive type is shown in the *Lepidosiren* (fig. 9), and the *Proteus* (fig. 10). The modification of the pelvic segments and their appendages in the reptile forms another prominent feature of deviation from the archetype. The pleurapophyses are continued, progressively shortening, attached to the diapophyses of a certain number of the vertebræ that succeed the sacrum : the hæmapophyses are no longer attached to their extremities, but are directly articulated to the central elements, with a slight degree of displacement, whereby they articulate to another segment as well as to their own. The mode and degree of departure from the archetype are now such that different series of vertebral segments may be classed into groups, with distinctive characters and names :—

The first four segments, by the fixed union of their neural arches, as *cranial* (Cr), under the collective name of ' skull.'

The next nine segments, moveably articulated, and with free or ' floating' pleurapophyses, as *cervical*, C, forming collectively the region called ' neck.'

The succeeding nine segments with ossified and moveable pleurapophyses and hæmapophyses, as *dorsal*, D, forming the ' back,' ' thorax' or ' chest.'

The three following moveable vertebræ, without free bony pleurapophyses, as *lumbar*, L, forming the ' loins.'

The next two vertebræ, immoveably united, and with modified and much-developed hæmal arches and appendages, are called *sacral*, and collectively ' pelvis and hind-limbs.'

All the other segments are ' *caudal* ' and constitute the ' tail.'

The hæmal arch (51, 52) with the developed appendages (53–57*a*) detached from the occipital vertebra, may require to be specially noticed in this summary of the parts of the endoskeleton, as from the circumstance of its commonly remote position from its proper segment, it may not have been thought of as a part included in the first class of vertebræ constituting the skull.

Many striking and extreme deviations from the archetype are manifested in the skeleton of the more aberrant forms of the reptilia. The number of moveable trunk-segments is reduced to the minimum in the *Batrachia* (*e.g.* 7 in *Pipa*), and increased to the maximum in the *Ophidia* (422 in *Python*). At first view the principle of vegetative repetition seems to have exhausted itself in the long succession of incomplete vertebræ which support the trunk of the great constrictors : but by the endless combinations and adjustments of the inflections of their long spine, the absence of locomotive extremities is so compensated that the degraded and mutilated serpent can overreach and overcome animals of far higher organization than itself : it can outswim the fish, outrun the rat, outclimb the monkey, and outwrestle the tiger ; crushing the carcase of the great carnivore in the embrace of its redoubled coils, and proving the simple vertebral column to be more effectual in the struggle than the most strongly developed fore-limbs with all their exquisite rotatory mechanism for the effective and varied application of the heavy and formidably-armed paws. And whilst the vertebral column of the ophidian order exhibits the extreme of flexibility, that of the chelonia manifests the opposite extreme of rigidity : back, loins and pelvis constitute one vast sacrum, or rather abdominal skull, but a skull subordinated chiefly to the lodgement and defence of a much-developed hæmal system, and in which the pleurapophyses, hæmapophyses and their spines appear to repeat the same modification of great expansion and fixed union by marginal sutures, which the neurapophyses and spines undergo in the cranium of the higher vertebrates. This formation of the ' carapace' is actually due, however, to the connation of dermal bones with the above-named vertebral elements, but in which it is worthy of note that the neurapophyses exhibit the modification of change of position, like that which has been described in the sacrum of the bird ; being shifted from their own centrum over one half of the next centrum, thus adding to the strength and elasticity of the whole osseous vault. The plastron is formed by the connation of dermal plates with the sternum and abdominal ribs. The confluence of the neurapophysis (14) with its own moiety of the neural spine (15) is characteristic of the anterior segment of the cephalic skull of most Chelonia. I may here add, that the typical condition of the hæmal (maxillary) arch of the same segment is well shown in the *Emys expansa*. The pleurapophyses (palatines) meet at the base of the cartilaginous vomer, above and behind the posterior nares, sweep outwards and downwards, give attachment to the hæmapophyses (maxillaries) which advance and converge, and the arch is closed below the nasal passage by the hæmal spine (premaxillary). Cut through the junction of the hæmapophyses with the neurapophyses (prefrontals), and with the diverging ap-

H

pendages (malars), and the inverted arch is then suspended by its proper piers, the pleurapophyses or palatines.

In the connation or coalescence of the neurapophyses and spines forming the parietal and frontal neural arches in the ophidian and some chelonian reptiles, we perceive a return to the common constitution of those arches in the vertebræ of the trunk, in which the permanent separation of the neural spine from the neurapophyses occurs as a rare exception.

In the class-skeleton (*Aves*) represented in fig. 4, the archetype is further departed from than in the typical Reptilia ; and when the general form of this diagram is contrasted with that of the first figure, the power of demonstrating the fundamental agreement which reigns throughout, and which is equally manifested in the comparison of figure 4 with those of the piscine and reptilian skeletons, affords a most striking proof of the unity of plan which pervades the whole series.

As compared with the crocodile (fig. 3), the proportions of the hæmapophysis (21) and spine (22) of the anterior segment are reversed ; there is a return towards the condition of the parts in fishes (fig. 2) ; the strength of the arch being chiefly due to the great development and extensive connections of 22, which usually sends a process upwards and backwards between the divided neural spine (15) of its own to that (11) of the next segment. The pleurapophysis (20) has often a slender rib-like form, and the appendages retain the form of bony rays. That (24) from the pleurapophysis is simple ; that (26, 27) from the hæmapophysis is divided in the embryo-bird : both concur in attaching the hæmal arch of the anterior segment to the pleurapophysis of the second segment. The neurapophyses of the anterior segment coalesce and form a single vertical bone, slightly expanded above and sometimes appearing anterior to the frontal.

The inferior or cortical part of the centrum, 9, of the second or frontal vertebra is connate, as in the fish (fig. 2, 9), with that of the third, 5 ; and the superior dismemberment of the second centrum, answering to 9' in fig. 2, is connate with the coalesced neurapophyses, 10.

The hæmal arch of the second segment is detached from its neural arch ; and, although its proper parapophysis (12) sometimes joins the next one (8), yet this exclusively supports, in birds, the pleurapophysis (28) of the frontal segment. The hæmapophysis is developed, as in the reptile, from several centres (29, 29', 30', 31), but these coalesce with each other and with the hæmal spine, 32, to form the single bone called lower jaw in most birds.

The hæmal arch (40–46) properly appertaining to 8—the parapophysis of the parietal segment—is detached from it, and freely suspended, somewhat retrograded in position beneath the next segment : its development has suffered as marked an arrest as in the crocodile.

The hæmal arch, with its appendages of the hindmost segment of the skull, is displaced backwards to a greater extent than in the reptile*.

The pleurapophysis (51) retains the form of a long, flat, slightly-arched rib : the hæmapophysis (52) is straighter and stronger. There are birds (*Apteryx*, e. g., fig. 9, p. 47) in which this arch is arrested at almost as early a stage of growth as is the antecedent (hyoid) arch of the skull. The elements of the neural arches of the skull, 1–15, early anchylose together in most birds, with the exception of the centrum (13) of the foremost segment, which more commonly coalesces with the pleurapophyses (20) of its hæmal arch.

The size of the brain now demands a modification of the neural arches superadded to those which they present in the cold-blooded vertebrates, and occasions a marked difference in the form of the skull : it is important to note how this is obtained. The nature of the modification is well shown in the young of those large birds which are devoid of the powers of flight. No new bone is introduced to increase the cranial walls and give the cavity its due

---

* The process of retrogradation is associated with the difference of time in the order of development of different segments of the endoskeleton. The typical vertebræ are usually the first to be completed, *e. g.* those of the thorax and head; the intervening ones are later developments, and the head is removed from the chest to an extent corresponding with their number or size. But the hæmal arch of the last cranial vertebra being detached in air-breathing Vertebrates from the rest of its segment, follows the heart, which it was primarily developed to support and protect, and consequently maintains its primitive contiguity with the thorax. The original relation of its hæmapophyses —the coracoids—to the pericardium is most instructively maintained in the Chelonia. In the higher air-breathing animals the proper thoracic hæmal arches take this function, and leave the scapular arch to subserve the offices of its greatly and variously developed appendages.

H 2

capacity ; this is gained by excess of growth of common and constant elements ; and those (3, 7, 11) furthest from the centrums are the chief seat of such excess. With regard to the neural spines of the frontal and parietal vertebræ, it is accompanied by a temporary bipartition, the ossification commencing from two lateral centres in each ; but the halves soon coalesce with each other and with their sustaining neurapophyses (2, 6, 10).

In those segments which, from the brevity and free termination of the pleurapophyses, may be called 'cervical,' the elements of the neural arch and also the pleurapophyses early anchylose together in each segment, converting it into the single bone, called in comparative osteology a 'vertebra,' and these vertebræ are remarkable for their great number in most birds ; and consequently the neck is as remarkable for its great length and flexibility. The detached hæmapophyses (58) of one of these vertebræ, (which vertebra, by the analogy of the fish (fig. 2, 58), should be the atlas,) commonly coalesced together at their distal ends forming a bony arch, like a slender edentulous lower jaw, have followed the hæmal arch of the occipital vertebra (51, 55) in its retrograde course, though not quite to the same extent. These mutually anchylosed hæmapophyses (58) forming the bone, called 'furculum' in ornithotomy, are generally the only pair of ossified cervical hæmapophyses. If, however, we define the cervical vertebræ, as in the crocodile, by their mobility and the free termination of their pleurapophyses, we may then recognise in some birds the detached hæmapophyses of the last cervical vertebra attached, as at h, to those of the succeeding segment : this structure may be observed in the common goose (*Anser palustris*). The pleurapophyses of the posterior cervical vertebræ are free, and rapidly elongate. The hæmapophyses of the segments with complete hæmal arches are bony, and are commonly defined as 'sternal ribs,' their pleurapophyses being called 'vertebral ribs,' agreeably with the restricted anthropotomical meaning of the term 'vertebra.' These pleurapophyses support bony appendages (a a), which serve, like those of the foremost hæmal arch of the skull, to connect their own arch to the next and associate them together in movement*.

* These appendages are not the result, as has been supposed, of a bifurcation of the vertebral rib : they are independent pieces originally in all birds, and retain their individuality in some, e. g. apteryx, penguin, with proper muscles for their elevation and depression—potential homotypes of the flexors and extensors of more developed limbs.

After six or seven segments with these typical hæmal arches come others with shorter pleurapophyses terminating freely, not reaching their hæmapophyses, one of which, ossified, is shown in the diagram at h', adhering by its distal end to the preceding hæmapophysis and terminating freely above. These 'floating sternal ribs' are more numerous in the crocodile (fig. 3, h'). The hæmal spines of the dorsal segments with complete hæmal arches become the seat of the most extensive and characteristic modifications of the avian type of skeleton. They are greatly extended in breadth, and, like the correspondingly expanded neural spines of the cranial vertebræ, are developed from two lateral moieties; but the individual spines, indicated by dotted lines in the diagram (60), are not ossified from separate centres, but continuously, so that the hæmal spines of six or eight vertebræ are at first represented by a pair of osseous plates. A cartilage is usually extended vertically from their median junction, which, when ossified, forms a strong crest or 'keel' (60'). The hæmal spine of the scapular arch (52') is sometimes ossified from a proper centre ; as is also a piece prolonging the series posteriorly : but all soon coalesce into one bone called 'sternum.' The anterior portion, 52', has received the name of 'episternum,' the median keel, 60', that of 'entosternum,' the posterior piece, which sometimes remains cartilaginous, that of 'xiphisternum.' In the terrestrial birds incapable of flight, the keel or 'entosternum' is not developed : in the rest of the class the extent of this part and of the ossified portion of the body of the sternum bears a direct ratio to their power of flight ; the peculiar modification of these extreme elements of the dorsal segments being governed by the size of the muscles moving the wings.

The next great deviation from the typical standard, peculiar to birds, is the great extent of the vertebral axis which is embraced by the enormously developed pelvic pleurapophyses, 62, and the unusual number of segments which, being thus deprived of reciprocal motion, grow together and form, according to this character, the bone or region called 'sacrum.' In investigating the structure of this part of the endoskeleton in the embryo-bird, the neural arches are found to manifest a change of position analogous to, though less extensive than, that of certain of the hæmal arches of more anterior segments (51–52, $e.\ g.$) : the results of this analysis are detailed at p. 61. Most of the pleurapophyses of the sacral vertebræ are stunted in their growth, which may literally be said to be stopped by the pressure upon their extremities of the overgrown distal portion of one of their homotypes, forming the bone called 'ilium' (62, *pl*). But one or two of the pleurapophyses at the anterior part of the series (*pl*) escape from beneath the 'ilium' to terminate freely at some distance below it: these are usually bifurcate at their proximal ends, and moveably articulated to their anchylosed centrums and diapophyses : the shorter anchylosed sacral pleurapophyses have simple proximal ends and articulate in the embryo to the interspace between their own and the adjoining centrum, as shown in fig. 10, *pl*, p. 61, to which they soon become anchylosed.

The contemplation of the modifications of the different natural segments in the trunk of the bird, particularly the freedom of some elements and the fixation of others, strongly impresses on the mind the purely artificial character of the regions of the spine which have been transferred from anthropotomy into the anatomy of the vertebrate animals. Thus Cuvier declares, " Il n'y a point de vertèbres lombaires proprement dites *." And a later author :—
" Die Wirbel zerfallen in Hals- Rücken- Kreuzbein- und Stusswirbel ; eigentliche Lendenwirbel sind gewöhnlich nicht zu unterscheiden."

Cuvier's negation in 1799 of proper lumbar vertebræ in birds is reproduced in succeeding systems and handbooks of comparative anatomy down to the latest by Siebold and Stannius, *e. g.* of 1846. But the student of anatomy in its wider acceptation will understand that the segments homologous with those included under L in fig. 3 are by no means wanting in fig. 4, but only otherwise modified.

It may be regarded as highly probable at least, from the striking points of agreement which are observable in the organization of the crocodile and of the bird, that, counting backwards from the first 'dorsal' in figs. 3 and 4, the next twenty segments are homologous in both. But, in the bird, those that answer to the three or four last dorsal vertebræ in the crocodile are anchylosed together, and the last of these had its pleurapophyses modified to form abutments against the elongated ilia. The next three segments, answering to the lumbar in the crocodile, are modified as in the last 'dorsal.' The two following segments similarly modi-

* Cuvier, Leçons d'Anatomie Comparée, i. (Ed. 1799, p. 170 ; Ed. 1836, p. 205).

fied will answer then to the two sacral vertebræ of the crocodile, and anchylosis extends backwards so as to include two or three vertebræ homologous with the anterior caudals in the crocodile. This appears to be the true interpretation of the enormous ' sacrum' of the bird; it is not merely ' lumbo-sacral' but ' dorso-lumbo-sacro-caudal,' including as it does representatives of each of those classes of vertebræ in the crocodile, but which have lost the artificial characters that distinguished them in that nearest allied existing vertebrate. The special homologies are indicated by the letters D, L and S.

The characters of the regions of the vertebrate skeleton are, as already remarked in reference to the crocodile, artificial, and are used for the sake of convenience in describing and comparing the vertebræ of different species. Those, therefore, are the best which are the most constant and most readily applicable in any given class. Proceeding to assign such to the bird, as in the crocodile, unbiased by anthropotomical characters of the vertebral regions, all those may be called ' cervical' in the bird that extend from the skull to the first vertebra with the hæmal arch complete, and those dorsal that extend from that vertebra inclusive, to the first vertebra embraced by, and anchylosed to, the iliac bones. One usually finds in the falcons, the gallinaceous birds and in some waders, five or six of the centrums and neural arches of the dorsal vertebræ anchylosed into one mass, a single free centrum usually intervening between this mass and the true sacrum. Some comparative anatomists call that cervical vertebra the ' first dorsal' in which the pleurapophyses retain, or begin to regain, their moveable articulations : but this character varies in different individuals of the same species. I have even found the pleurapophysis of the last cervical vertebra anchylosed on one side and not on the other.

The retention by the pleurapophyses of moveable articulations with the centrum, might also seem a good character of dorsal vertebræ at the hinder end of the series ; but it is inconstant : I have found those elements anchylosed in one individual and free in another of the same species, in the anterior vertebræ, which are sacral by coalescence. All those vertebræ may be called for convenience ' sacral' in the bird, which are confluent by both centrums and neural arches with each other and with the iliac bones; and this confluence is so complete that it usually requires a vertical section and reference to the nerve-outlets in order to determine their number. The free vertebræ that succeed these are the caudal, of which the last, as in most osseous fishes, is a coalesced congeries of several, though for convenience counted as one, and called in ornithotomy the plough-share bone (*c, n, h*). Although so many segments of the bird's skeleton are modified to transfer the weight of the horizontal trunk upon the ilia (62), the 'pelvis,' as in the crocodile, has but two hæmapophyses, 63, 64, below : it is charac-teristic of birds, however, that these are not united at their distal ends to their fellows of the opposite side, either with or without the intervention of a hæmal spine. The exception which the ostrich offers in regard to the anterior pair (pubic bone, 64) and that which the rhea presents in respect to the posterior pair (ischia, 63), serve to prove the rule of the inferiorly open pelvis in birds.

In regard to the diverging appendages of the two hæmal arches (scapular and pelvic) which have been selected for development into locomotive organs in all classes of vertebrata, the corresponding segments (carpal, 56, and tarsal, 68) agree in the paucity of their divisions, two bones in each, in all birds ; and the succeeding seg-ments (metacarpal and metatarsal) in consisting of three coalesced bones in both wing and leg, supporting digits answering to those marked II, III and IV, *ii, iii, iv*, in the crocodile. Such at least is their general character, the minor differences being the follow-ing :—In the hand-segment of the wing the anchylosed metacarpal of digit II is very short, represented as it seems only by its proxi-mal end ; those of the digits numbered III and IV attain their normal length, and are anchylosed together at the extremities only, with an interspace between their shafts.

The wing of the bird is chiefly formed by the quill-feathers, developments of the hard, insensible epidermal system : and this expanse forms a striking contrast with the delicately organized and highly sensitive web which is stretched over the long and slender finger-bones of the bat. Its period of activity in the hours of gloom necessitated the accessory sense of fine touch in aid of vision to safely guide its flight.

In the metatarsus of the bird the three bones coalesce through-out their length, except in the penguin, where interspaces are left between their shafts or middle parts. But they also coalesce proximally with the two primitively distinct tarsal bones (68), whilst the metacarpals coalesce proximally with only part of the carpal series, if at all. And to the metatarsus there is usually superadded a rudimental, but unanchylosed, metatarsal bone of the digit answering to no. 1 in the crocodile ; but directed backwards, except in the swift. The numbers of the phalanges of the toes, *i*, *ii* and *iii* in birds, correspond with those of their homologues in the crocodile : the toe *iv* has an additional phalanx, and the regular progression of the increase from two to five phalanges, with one or two exceptions, is constant in the class, and serves to determine the toes in those birds in which they are reduced to three or two in number. Thus in the ostrich (fig. 11), the shorter of the two toes is determined by its greater number of phalanges, 5, to be the homologue of the fourth in tetradactyle birds ; and it is interesting to observe that the toe *iii*, notwithstanding its much greater length, has the usual smaller number of phalanges. But whilst unity of design is thus manifested, the wisdom of the Designer is displayed by the greater strength which results from the minor degree of subdivision of the part which takes the largest share in the support and propulsion of the body. The toe *v* is never present in birds ; there is not even the rudiment of its metatarsal bone. The toe *i* is equally absent. (See paragraph at p. 36, on the spurs of the *Gallinacea*.)

Fig. 4 is the diagram of the skeleton of a typical mammalian quadruped ; *e. g.* the dog (genus *Canis*). The modifications of the hæmal arch of the anterior segment resume the characters of those in the crocodile ; the hæmapophysis (21) being the chief seat of development, and for the same purpose of extending its attachments, and adding to the firmness and strength of the henceforth immoveable maxillary arch. The diverging appendage from the pleurapophysis (20) is a single bone on each side (24), and in most mammals becomes confluent with the part of the posterior segment (5) against which it abuts.

The neurapophyses (14) of the anterior segment have coalesced together, as in birds, but are complicated and their nature further obscured by anchylosis with ossified portions of the olfactory capsules, often extremely complex and extensive in the class Mammalia, in which the organ of smell attains its maximum of development. The neural spine (15), sometimes single, more frequently bifid, enjoys, agreeably with its extreme position in the series, a vast range of variety in its forms and proportions. In the rhinoceros it supports a dermal spine or ' horn.'

The second (frontal) segment presents unexpectedly a return to the archetypal character in a particular, in the absence of which all the lower classes of vertebrata depart from it, viz. the primitive independence of its centrum (9) from that (5) of the succeeding segment. The spine (11) of this, as well as those (7, 3) of the two following segments, continue, as in birds, to be the chief seat of the expansion requisite for the protection of the progressively developing brain. But in most mammals an additional element in the cranial walls is gained by the expansion of the distal end of the diverging appendage from the hæmopophysis (21) of the anterior segment. This appendage consists, as in birds and reptiles, of two pieces, and it is the second or most remote piece (27) which is the seat of the principal varieties, and especially of that squamous development which enables it not only to extend the points of fixation of the maxillary arch, but at the same time to subserve the requirements of cranial space consequent on the large size of the cerebrum. The dismemberment called 'interparietale,' x, of the spine 3, has a less constant relation to the increased capacity of the cranium.

The pleurapophysis (28) of the second segment becomes, in the present class, still further displaced from its typical connections, and is even superseded in its typical functions by the intervention and development of 27. It is consequently much reduced in size, and strangely distorted in form, in subserviency to the almost sole office that now remains to it, viz. the support of the tympanic membrane.

The frontal hæmapophyses and spine (29–32) have coalesced into a single bony arch, articulated by its extremities to the under part of the appendage 27.

The pleurapophysis (38) of the hyoid or third hæmal arch resumes in many mammals its typical connections with the parapophysis (8) of its proper segment; but its development is usually more or less restricted.

The articulation of the fourth (occipital) segment with the succeeding one called 'atlas,' is chiefly by means of zygapophyses (condyles) developed from the neurapophyses (2); the parapophyses (4) are likewise exogenous processes of the same elements.

The hæmal arch of the occiput (51, 52), though in close proximity with its proper neural arch in some mammals, and in all mammals at the earlier period of development, is not directly articulated thereto, and sometimes recedes far from the rest of the skull.

The hæmapophysis (52) of the arch is ossified throughout its entire extent and articulated with the hæmal spine 52', below, in only one small exceptional order of the class (*Monotremata*). It becomes anchylosed with the pleurapophysis in all, and appears in the majority therefore as a mere process of 51.

The single pair of cervical hæmapophyses (58) are more variable, both as to their extent of ossification and even existence.

The body of the atlas continues subject to the same modification of development from two centres with coalescence of one portion with the next centrum, which characterises it in all the other vertebrates above Batrachians*.

The confluence of the centrum with the neural arch takes place in every vertebra of the trunk; and the pleurapophyses, which are very short in the seven segments that succeed the skull, here also commonly coalesce with the other elements, circumscribing the lateral foramina for the 'vertebral' arteries. With the exception of the detached bones 5s, they are the only ossified parts of the hæmal arches of those segments.

The constancy of the number, *seven*, of the segments so modified, is truly remarkable and characteristic of the class Mammalia. It is true that the number is established at a very early stage of development, when the neck is alike short in all; and its law must be sought for in the circumstances, such as the existence of a complete diaphragm in the mammalia, which determined the number and distribution of the pairs of cervical nerves, upon which the development of the cervical vertebræ more immediately depends. The exceptions to the number *seven*, viz. *six* in the manatee, and *eight* or *nine* in the three-toed sloths, serve to establish the rule.

The eighth segment of the trunk in mammalia, like the tenth in the crocodile, has a complete hæmal arch, and here therefore the 'dorsal' series begins; but the hæmapophysial elements are rarely ossified in the present warm-blooded class.

The pleurapophyses (*pl*) of these arches are not only moveable, but are subject to a slight displacement, and their articulations, like those of the neurapophyses in the bird's sacrum, extend over the interspace of their own and a contiguous centrum.

* See Taylor's Annals and Magazine of Natural History, vol. xx. p. 217.

The hæmal spines (60, 61, *hs*) commonly remain distinct, and form a chain of ossicles corresponding in number with the complete hæmal arches, but they coalesce with each other in some of the higher mammalia, and are called collectively 'sternum.'

As the segments recede the pleurapophyses become shorter, return to their proper vertebra, and usually become appended to its diapophyses; the hæmapophyses also become shorter, and terminate at first by abutting against their antecedents, and finally by projecting freely.

These segments are followed by others (L) in which only the pleurapophysial parts (*pl*) of the hæmal arch are ossified, and these parts coalesce with the diapophyses (*d*).

Then come the segments (S), which, like those of the skull, are the seat of the modification by anchylosis, and of great and peculiar development of two of the hæmal arches in connection with them; the nature of the deviations from the typical standard which characterise the province of the endoskeleton called 'sacrum' and 'pelvis,' has been explained at pp. 61 & 73. In most mammals a greater number of segments is involved in this metamorphosis than in reptiles; in none are so many the seat of it as in birds. In the cetacea the modification by anchylosis is transferred to segments at the fore-part of the trunk; their 'sacrum' may be said to be in the neck; none of the post-abdominal vertebræ are subject to confluence any more than in serpents, fishes, or the extinct marine reptiles (*Enaliosauria*).

Great diversity of form, of number and of development prevails in the vertebræ that succeed the sacrum in mammalia. Short pleurapophyses are developed at the extremities of the diapophyses of the anterior ones and coalesce with them. The hæmapophyses, when present or ossified, are articulated, as in reptiles, to the centrum directly, and alone form the hæmal arch. The terminal vertebræ are reduced to the central element, and rarely anchylose together.

The anterior anchylosed and expanded vertebræ are the cranial, Cr.

Those usually free vertebræ with short pleurapophyses, anchylosed to both their centrum and neural arch, are called 'cervical,' C. In some whales and armadilloes all or some of these vertebræ coalesce into one mass.

The series with moveable and usually longer pleurapophyses is called 'dorsal,' D.

Those with pleurapophyses confluent or connate with the extremities of the diapophyses are called ' lumbar,' L.

The succeeding vertebræ which anchylose together are called ' sacral,' S.

The rest are ' caudal,' Cd.

The modifications of the diverging appendages of the scapular and pelvic arches are numerous in kind and extreme in degree : with the exception of the cetacea, in which the hinder pair is absent—the cheiroptera, in which the fore-pair is specially developed for the actions of flight—and some burrowers, as the mole—a close analogy is commonly kept up between the two pairs : both, for example, are reduced to the same degree of simplicity in the solidungulous horse ; both arrive at almost the highest stage of development, in the special adaptation of one of the digits to react upon the rest as an opposable thumb in both the fore- and hind-feet of the Quadrumana.

Fig. 15, bones of the fore-limb, and fig. 16, bones of the hind-limb, of the Wombat, illustrate the serial homology of those bones, explained at pp. 23 & 24.

53, 'humerus,' is the homotype of 65, 'femur.'

54, ' ulna,' is the homotype of 67, ' fibula.'

o, ' its olecranon,' is the homotype of 67', ' fabella,' or the sesamoid bone articulated to the produced and expanded head of the fibula.

55, 'radius,' is the homotype of 66, ' tibia ' *.

sc, scaphoid portion of ' os scapholunare,' is the homotype of sc, ' scaphoides.'

l, lunar portion of ' os scapholunare,' do. of a, ' astragalus.'

cu, cuneiform portion of ' os scapholunare,' do. of cl, articular part of ' calcaneum.'

p, ' pisiforme,' is the homotype of cl', fulcral part of ' calcaneum,'

t, ' trapezium '	do.	of ci, inner cuneiform.
z, ' trapezoides '	do.	of cm, middle cuneiform.
m, ' magnum '	do.	of ce, outer cuneiform.
u, ' unciforme '	do.	of b, cuboides ; both of these

representing two distinct carpals coalesced, as the scapholunar

---

* The tendon of the triceps femoris is not ossified in this species, where it passes over the knee-joint at 66' ; it resembles in this respect its homotype, the tendon of the biceps brachii, in the fore-limb.

in the carpus represents the astragalus and scaphoid in the tarsus, and the calcaneum reciprocally the cuneiform and pisiform bones.

The serial homologies of the carpals and tarsals are better illustrated in the hand (fig. 13) and foot (fig. 14) of the orang, as has been before explained.

Having thus noticed some of the chief varieties of the mammalian modification of the vertebrate archetype, there remains to add only a few words in explanation of fig. 6,—the diagram of the human skeleton.

As this is that which the anatomist has been accustomed to hear described most frequently and exclusively by the special terms, and according to the special views and ends of anthropotomy, the language in which its deviations from the common archetype have now to be noticed will probably appear strange and bizarre. The comprehension of the explanation will be facilitated by reference to the special name of the bone through its numeral in the column of names whenever such bone is alluded to under its general or archetypal name.

In the first and, notwithstanding the upright posture, the most anterior of the cranial segments, by reason of their forward curvature, the hæmapophysis (21) coalesces early with its own moiety of the divided spine (22), and the same thing happens to the next hæmal arch (29) with subsequent obliteration of the symphysis between the halves of its spine (32).

The pleurapophysis (20) of the first arch remains a distinct bone : its diverging appendage (24) coalesces with and becomes a 'process' of the centrum (5) of the parietal vertebra.

The neurapophyses (14) of the anterior segment are modified as in other mammalia, i. e. become confluent together and with the olfactory capsules ; but appear externally below the orbital process of the frontal.

The spine (15) is small, but bifid.

That of the second segment (11) attains its maximum of development, as do also the spines of the two following vertebræ (7 and 3). The bifid spine of the parietal segment is truly enormous as compared with that of the fish (fig. 2, 7) or the reptile (fig. 3, 7), in which latter animal the spine, being undivided, adheres closer to the archetype.

The diverging appendage (26, 27) from the hæmapophysis (21) is divided into two pieces, as in most mammals and reptiles ; both are broad and flat : the first (26) serves to fix the arch to the parapophysis (12) of the second segment, from which it is here dislocated ; the portion (27), which becomes enormously expanded, covers a large vacuity between the third and fourth neural arches, and overlaps by a squamous suture part of the expanded spines of both those vertebræ. It also anchyloses below with the pleurapophysis (28) of the second segment, with the parapophysis (s) and the pleurapophysis (38) of the third segment, as well as with the bony capsule of the organ of hearing (16), forming with those parts the most singularly complex ' cranial bone '—the ' os temporis' of anthropotomy.

The centrums (5, 9) and neurapophyses (6, 10) of the second and third segments coalesce with each other, and with the first pair of diverging appendages (24) of the anterior hæmal arch (20, 21, 22), forming the complex ' sphenoid' bone of anthropotomy.

The centrum (1), neurapophyses (2), and neural spine (3) of the fourth segment speedily anchylose together, and their centrum afterwards coalesces with that (5) of the parietal vertebra, forming the still more complex cranial bone called ' os spheno-occipitale' by Soemmering.

The hæmapophyses of the third much-reduced hæmal arch (40) are ossified only at the extremity which joins the spine (41) : the remainder of the hæmapophysis is continued in a ligamentous state to their anchylosed pleurapophyses (38), forming the ' styloid processes of the temporal bone.'

The detached and displaced pleurapophyses (51) of the occipital vertebra attain considerable breadth : their hæmapophyses (52) are ossified only at the extremity which joins the pleurapophysis, and with which it coalesces. The diverging appendage (53–57) here attains its maximum of adaptive development ; as in the skate-fish (*Raia*) it exhibits the extreme of vegetative or polaric growth. But the progressive steps by which it departs from the primitive or archetypal simplicity, shown in figures 7 and 8, are so gradual, that the special homology of the arm and hand of man with the bifid-jointed appendage of the scapular arch in the amphiuma, and with the simple jointed ray of that of the scapular arch of the lepidosiren, has never been doubted or called in question. In ascending, therefore, to the higher generalization of the signification, or relation to the archetype, of such simple, or bifid, jointed or more complicated appendage of such scapular arch, we are compelled by the truth, as it exists in nature, to admit that the scapular arch in the lepidosiren and other fishes forms the inferior costal or hæmal arch of the occipital segment or vertebra; and, by reference to the archetype, to see in the diverging appendage of such arch, a repetition of similarly simple diverging appendages of succeeding segments. These, indeed, retain their primitive simplicity, as shown in the trunk-vertebræ of the fish (fig. 2, *a a*) and of the bird (fig. 4, *a a*); and that simplicity is very gradually departed from in the case of the appendages of the occipital vertebra, by the stages recognisable in figs. 7 and 8. If, then, it be admitted that the upper limb (arm and hand) of man is the homologue of the fore-limb of the amphiume, of the pectoral fin of the fish and of the pectoral ray of the lepidosiren; it follows, that, like the latter, it must also be the 'diverging appendage' of the arch called 'scapular,' which is the hæmal arch of the occipital vertebra; and, therefore, however strange or paradoxical the proposition may sound, that the scapular arch and its appendages, down to the last phalanx of the little finger, are truly and essentially bones of the skull.

The centrum of the first segment of the neck is subject to the same modification as in the ordinary mammalia, the major part (*c a*) remaining anchylosed to the centrum of the succeeding segment (*c d*), of which it forms the 'odontoid process' in human anatomy. The inferior cortical part (*c, a, x*), or 'hypapophysis,' is that which is usually called the 'body' of the atlas: it is connected by aponeurosis to the corresponding part of the centrum of the occipital vertebra: the articulation of the head with the neck is chiefly by means of zygapophyses developed in the form of convex condyles from the neurapophyses (2), and received by the concave zygapophyses of the neural arch of the atlas. In the other cervical segments, the autogenous elements of which they are composed are represented diagrammatically in fig. 6 as distinct, viz. the centrum, neurapophysis, neural spine, and pleurapophysis; the latter element in the seventh vertebra sometimes attains a length nearly equal to that of the first dorsal. The hæmapophyses of the atlas (58: see also fig. 2, 58) are wholly ossified and well-developed.

In the seven vertebræ which succeed the cervicals the pleurapophyses (*pl*) are progressively elongated; they are shifted from their proper centrum to the interspace between it and the next segment above, or in advance, and retain their moveable joints. The hæmapophyses (*h*) are cartilaginous and articulate with the ends of the pleurapophyses and with the hæmal spines (*hs*), which are flattened, slightly expanded, and ultimately blended into one bone called 'sternum.' The hæmal spine of the first thoracic segment remains longest distinct: it receives, also, the extremities of the displaced hæmapophyses (53), and has been called 'manubrium sterni.' The hæmal spine of the seventh segment commonly continues longer distinct, and is later in becoming ossified, whence it is called 'ensiform cartilage': it probably includes the rudiments of some succeeding hæmal spines. In the four succeeding segments the pleurapophyses become progressively shorter, and the hæmapophyses, still cartilaginous, are severally attached by their lower attenuated ends to the pair in advance; leaving the hæmal arch incomplete below. In the next vertebra (19th from the skull) the still shorter pleurapophyses resume the exclusive articulation with their proper centrum; and the correspondingly short and pointed hæmapophyses are appended to their extremities and terminate freely.

Those pleurapophyses and hæmapophyses which directly articulate with hæmal spines (sternum) are called collectively 'true ribs' (costæ veræ), the proximal element being 'the bony part of the rib' (pars ossea costæ), the distal one 'the cartilage of the rib.' The rest of the hæmal arches which are incomplete through the absence of the hæmal spine, are called 'false ribs' (costæ spuriæ); and the last, which terminates freely in the origin of the diaphragm, is a 'floating rib.' The centrum, neurapophyses, and neural spine of each segment with freely articulated pleurapophyses coalesce into one bone, called 'dorsal vertebra' in anthropotomy: these vertebræ are twelve in number. Each of the five succeeding segments is represented by the same elements (centrum and neural arch) coalesced that constitute the so-called dorsal vertebræ: they are called 'lumbar vertebræ': they have no distinct ossified pleurapophyses, but rudiments of such are connate with and lengthen out the diapophyses. The hæmal arches in the abdominal region retain their aponeurotic texture: the anchylosed and stunted pleurapophyses are continued by the tendinous origins of the 'transversus abdominis'; the hæmapophyses are the 'in-

I

scriptiones tendineæ recti abdominis ;' and the basis of the hæmal spines is the 'linea alba.' Certain elements of the five succeeding segments coalescing together in the progress of growth form the bone called 'sacrum,' and are described individually as sacral vertebræ. The first four of these each combine the same elements, coalesced, as in the neck ; viz. centrum, neurapophyses, neural spine, and short but thick pleurapophyses*. One typical segment, the second, is completed by the meeting of the broad sides of the inverted arch (62, 63, 64) at the 'ischio-pubic symphysis' forming the 'pelvis' of anthropotomy.

The first sacral vertebra has its hæmapophysis (64, *pubis*) ossified, but separated from its proper pleurapophysis by the expanded (iliac) portion of that of the succeeding vertebra, with which it coalesces, as well as with the hæmapophysis (ischium) of the same vertebra.

The second sacral vertebra has its pleurapophysis divided, and the lower portion expanded to form the so-called 'ilium' (62). The hæmapophysis (63) coalesces with that of the preceding vertebra (64), and with its own pleurapophysis (62, fig. 6).

The short and thick pleurapophyses of the third sacral vertebra also articulate in the adult with the expanded distal portions of those of the second sacral vertebra : but these (iliac bones) are restricted in infancy and early childhood to their connections with the first and second sacral vertebræ, which connections are permanent in most reptiles (fig. 3).

The fourth sacral vertebra consists of centrum, neurapophyses, and rudimental pleurapophyses ; the fifth sacral vertebra of centrum and rudimental neurapophyses, which rarely meet above the neural canal.

In each sacral vertebra the elements of the neural arch and rudimental ribs first coalesce together ; and afterwards the vertebræ unite with each other and form the anthropotomical bone called ' sacrum.'

The first coccygeal vertebra in man consists of a centrum and of stunted exogenous neurapophyses† wide apart above, but deve-

* J. Müller notices the rudimental ribs in the first and second sacral vertebræ of the human fœtus in his 'Anatomie der Myxinoiden,' heft i. 1834, p. 240. Mr. Carlile has described (Report of British Association, 1837, p. 112), and Dr. Knox has figured (Lancet, 1839, p. 191), these ribs and their homotypes in the third and fourth sacral vertebræ.

† " Shoulders of the os coccygis."—Monro, *l c.* p. 142.

loping zygapophyses, which join those of the last sacral vertebra, and diapophyses which extend outwards further than those of the same vertebra. The neurapophyses are represented by exogenous tubercles of bone in the second coccygeal vertebra ; and the third and fourth vertebræ are reduced to the centrums only.

The cartilaginous deposits in the primitive blastema of this extremity of the trunk indicate a greater number of caudal vertebræ, and the rudimental tail is proportionally longer in the embryo than in the adult. It is shortened, however, by absorption prior to the commencement of ossification, and but four segments are indicated by depositions of the earthy salts in the situations proper to the above-specified elements of a typical vertebra : these finally coalesce into a single bone " of a crooked pyramidal figure," which got its name of ' os coccygis ' from its supposed resemblance to a cuckoo's beak *.

The early recognition of these and other specialities arising out of the various adaptive modifications of the typical segments of the human skeleton found its expression, necessarily, in special terms, the convenience of which will ensure their permanence ; but the progress of anatomical science having unfolded the primary form which is the basis of those modifications, there arises the same necessity for giving utterance to ideas of the generic character of the parts by general terms.

Inasmuch, however, as the different segments of the human skeleton deviate in various degrees from the common archetype, and as the different elements of such segments differ in their modifiability, anthropotomy has at no period wanted also its ' general terms ' expressive of the recognised extent of such conformity : such terms also, indicating, obscurely indeed, so much perception of the pre-existing model as could be obtained from the study of one form, at a period when that form—the human frame —was viewed as something not only above, but distinct from, if not antithetical to the structures of the brute creation, and when it was little suspected that all the parts and organs of man had been sketched out, in anticipation, so to speak, in the inferior animals. Thus the word ' vertebra ' shows, by the number of the segments or parts of segments to which it is applied in anthropotomy, a recognition of the degree in which the principle of repetition of similar parts more obviously prevails in the construc-

* Monro, *l. c.* p. 141.

I 2

tion of the human endoskeleton. And, inasmuch as in some regions (the cervical, *e. g.*) the ' vertebra' includes all the elements of the typical segment, there developed, it has been retained in homological anatomy, but, with a more consistent and definite meaning, as the technical term of the primary segment of the endoskeleton in all vertebrate animals.

The ' true vertebræ' of anthropotomy are those segments which retain the power of moving upon each other ; and the term is applied in a peculiar and empirical sense very different from the meaning which the anatomist attaches to a true or typical vertebra. The ' false vertebræ ' of anthropotomy are those segments or parts of segments forming the lower or hinder extreme of the endoskeleton, and which do not admit of reciprocal motion at their joints. And Monro, admitting that the condition of even the human os coccygis sometimes militates against the definition, meets the objection by arguing for the speciality of that bone, and with as good or better reason than those who have subsequently contended against admitting the cranial segments into the category of vertebræ. " From the description of this bone " (os coccygis), " we see how little it resembles *vertebræ* ; since it seldom has processes, never has any cavity for the spinal marrow, nor holes for the passage of nerves*."

Embryology has since demonstrated that the parts of the os coccygis are originally in vertebral relation with the neural axis ; and that this is subsequently shortened by a concentrative movement, which in like manner withdraws it from the terminal segment at the opposite extreme of the endoskeleton. The homology of the divisions of the sacrum with the true vertebræ is admitted by Monro, because of the perforations for the nerves : and this character is still retained in the nasal vertebra in the form of the cribriform foramina, although its neurapophyses, like those of the sacrum, have lost their primitive relation to the neural axis.

Homological anatomy, therefore, teaches, that the term ' vertebra' should not only be applied to the segments of the human skeleton in the technical and definite sense illustrated by figs. 7 and 8, pp. 42 & 43, but be extended to those modified and reciprocally immoveable segments which terminate the endoskeleton superiorly, in Man, and are called collectively ' skull.' (figs. 1 to 6, Cr.)

The term ' head,' then, indicates a region of specially modified

---

* Monro, *l. c.* p. 143.

vertebræ, like the terms 'neck,' 'chest,' 'loins,' &c. ; and amongst the groups of the primary segments characterized by specific modifications, the 'cranial' vertebræ must be added to the 'cervical,' 'thoracic or dorsal,' 'lumbar,' 'sacral,' and 'coccygeal or caudal.'

Such, with reference to the 'general' term 'vertebra,' seems to be the advance of which anthropotomical science is susceptible, in order to keep pace and be in harmony with anatomy.

As to the elements of the typical vertebra, anthropotomy has also its general phrases (see Table II. column vi. 'Soemmering,' in my Work 'On the Archetype of the Vertebrate Skeleton'), some of which are equivalent to the clearly-defined technical terms of such elements in anatomy properly so called.

The serial homology of the centrum (*corpus vertebræ*) has been recognised in all the so-called 'true vertebræ,' and in some of the 'false vertebræ' : thus Monro says, " The fore-part of the *os sacrum*, analogous to the bodies of the true vertebræ, is smooth and flat*." But their smooth and flat homotypes in the skull have only the special names of 'basilar' and 'cuneiform' processes ; of 'processus azygos' and 'vomer.' The 'neurapophyses' are recognised as repetitions of the same part under the definitions of 'a bony bridge produced backwards from each side of the body of the vertebra,' of '*arcus posterior vertebræ*,' of 'vertebral laminæ' or 'pedicles.' Monro describes these rudimental elements in the last sacral vertebra, where they are 'exogenous,' as 'knobs,' and in the first coccygeal vertebra as its 'shoulders.' In the skull they receive the special definitions of " the pieces of the occipital bone situated on each side of the great foramen ; from which nearly the whole condyles are produced†" (*partes laterales seu condyloideæ*, Soem.) ; 'great' or 'temporal wings of the sphenoidal bone'‡ ; 'orbitar wings' or 'processes of the sphenoidal bone ;' 'nasal' or 'vertical plate' and 'crista galli' of the ethmoid ('*pars media ossis æthmoidei*,' Soem.).

The neural spines are called generally '*spinal processes*' in every segment of the trunk : in the head they are known only by the special names of 'occipital plate,' 'parietal bones,' 'frontal bone,' 'nasal bones.'

The pleurapophyses, when free, long, and slender, are called 'ribs,' 'vertebral ribs,' or 'bony parts of the ribs'; when short

* Monro, *l.c.* p. 138.　　　　† *L.c.* p. 76.　　　　‡ *L.c.* p. 86.

and anchylosed, they are called, in the neck, "the second transverse processes that come out from the sides of the body of each
vertebra*;" (*radix prior processus transversi vertebræ,* Soem. ;)
in the sacrum, 'transverse processes' and 'ilium'; in the skull,
'scapula,' 'styloid process of the temporal bone,' 'external
auditory or tympanic process of the same bone;' 'palatine bone.'

In like manner the serial homology of the hæmapophyses is
recognised in the thoracic region by the general term 'cartilages
of the ribs' or 'cartilages of the sternum'† there applied to the
same elements of twelve successive segments. When ossified in
other vertebræ they have received the special names of 'ischium,'
'pubis,' 'coracoid process of the scapula,' 'clavicle,' 'appendix
or lesser cornua of the hyoid bone,' ('*crura superiora,*' '*os lin-
guale superius,*' Soem.), 'lower jaw' or *mandibula,* 'upper jaw'
or *maxilla.*

The exigences of descriptive anthropotomy and its highly important applications to Medicine and Surgery necessitate such
special nomenclature, and the reform which that nomenclature
chiefly requires is the substitution of names in the place of phrases
for the parts of the human body.

But the retention and use of specific names for specially modified elements in the different segments by no means preclude the
entertainment of general ideas and the necessity of expressing
them by generic names for the homologous elements in the entire
series of vertebræ.

If anthropotomy is to make corresponding progress with anatomy, and to derive the same light from the generalizations of
zootomical science which medical botany has derived from general
botanical science, its nomenclature must expand to receive those
generic terms which express the essential nature of the parts, here-
tofore named and known only according to the results of parti-

---

* Monro, *l. c.* p. 126.

† Laurentius, in describing the human thoracic pleurapophyses, says,
"Earum duplex est articulatio, altera cum spondylis dorsi, altera cum sterni
cartilaginibus" (Anatomica Humani Corporis, Fol. 1600, p. 94). The perception of the essential distinctness of the vertebral ribs had not then been
blunted by the constant repetition of the conventional idea of their forming
an ossified part of a whole, *i. e.* a rib, completed by the hæmapophysis under
the name of the 'cartilago costæ.' In birds it is not uncommon to find the
hæmapophyses not only ossified, but some of them attached to the sternum,
and detached from the pleurapophyses.

cular and insulated observation. A term which truly expresses the general homology of a part enunciates the most important and constant characters of such part throughout the whole animal series, and implies therefore a knowledge of such characters in that part of the human body, when used and understood by the human anatomist. Before the cuneiform process of the occipital bone could be defined as the ' occipital centrum,' the modifications and relations of the homologous part in all classes of vertebrate animals had to be accurately determined. The generic homological term expresses the sum or result of such comparisons, and the use of such terms by the anthropotomist implies his knowledge of the ' idea' or primal pattern of the human frame supporting the modifications that raise it to an eminence so far above those of all other vertebrate animals.

In no species, however, is each segment of the endoskeleton so plainly impressed with its own individual characters, as in Man ; the practised anthropotomist, for example, will at once select and name any given vertebra from either the cervical, the dorsal, or the lumbar series. During that brilliant period of human anatomy which was illuminated by a Fabricius, an Eustachius, a Fallopius, and a Laurentius, the terms expressive of the recognition of such specific characters were more numerous and often more precise than in our modern compilations. Pleurapophyses were individualized in the thorax as well as in the head : the ' antistrophoi,' ' stereai' and ' sternitides,' for example, were distinguished from the other ' pleurai gnesiai ' *.

General anatomical science reveals the unity which pervades the diversity, and demonstrates the whole skeleton of man to be the harmonized sum of a series of essentially similar segments, although each segment differs from the other, and all vary from their archetype.

* Anatomica Humani Corporis, &c., multis controversiis et observationibus novis illustrata. Andr. Laurentio, fol. 1600, p. 95.

THE END.

Printed by Richard and John E. Taylor, Red Lion Court, Fleet Street.

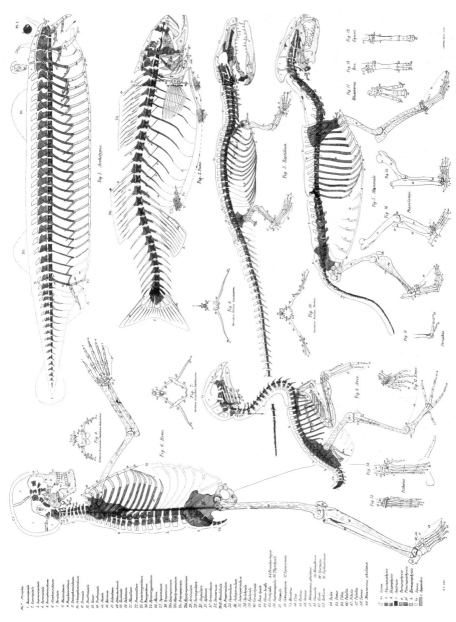

Plate 1. Full-size graphic is available at http://www.press.uchicago.edu/books/owen

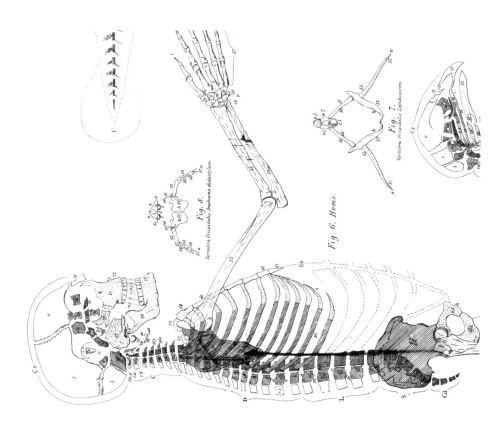

Fig. 8.

Vertebra Occipitalis Amphiuma didactylum.

Fig. 6. *Homo.*

Fig. 7.

Vertebra Occipitalis Lepidosiren.

Nu.ᵃ	Nomina.
1.	Basioccipitale
2.	Exoccipitale
3.	Supraoccipitale
4.	Paroccipitale
5.	Basisphenoideum
6.	Alisphenoideum
7.	Parietale
8.	Mastoideum
9.	Prosphenoideum
10.	Orbitosphenoideum
11.	Frontale
12.	PostFrontale
13.	Vomer
14.	PreFrontale
15.	Nasale
16.	Rerosum
17.	Sclaroticum
18.	Ethmoturbinale
19.	Turbinale
20.	Palatinum
21.	Maxillare
22.	Premaxillare
23.	Entopterygoideum
24.	Pterygoideum
25.	Ectopterygoideum
26.	Malare
27.	Squamosum
28.	Tympanicum
28a.	Epitympanicum
28b.	Mesotympanicum
28c.	Pretympanicum
28d.	Hypotympanicum
29.	Articulare
29.	Surangulare
30.	Angulare
31.	Spleniam
31ᵗ.	Coronoideum
32.	Dentarium
29.32.	Mandibula
34.	Preoperculum
35.	Operculum
36.	Suboperculum

Plate 1, detail 1

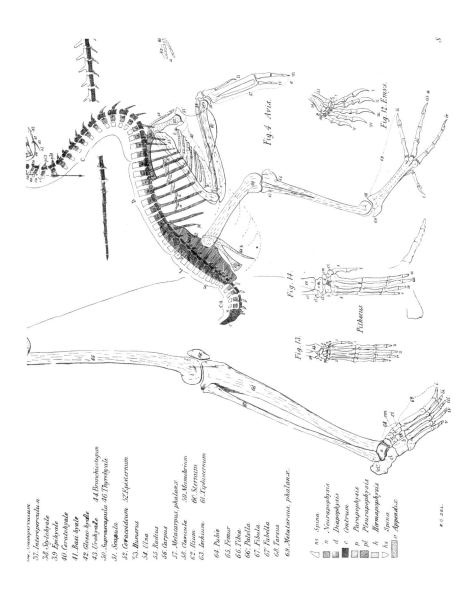

Plate 1, detail 2

36. Suboperculum
37. Interoperculum
38. Stylohyale
39. Epihyale
40. Ceratohyale
41. Basi-hyale
42. Glosso-hyale
43. Urohyale   44. Bronchiostegum
50. Suprascapula 46. Thyrohyale
51. Scapula.
52. Coracoideum   52. Episternum
53. Humerus
54. Ulna
55. Radius
56. Carpus
57. Metacarpus, phalanx
58. Clavicula   50. Manubrium
62. Ilium   60. Sternum
63. Ischium   61. Xiphisternum
64. Pubis
65. Femur
66. Tibia
66. Patella
67. Fibula
67. Fabella
68. Tarsus
69. Metatarsus, phalanx.

ns   Spina.
n   Neurapophysis
d   Diapophysis
c   Centrum
p   Parapophysis
pl   Pleurapophysis
h   Hæmapophysis
hs   Spina
a   Appendix

Fig. 4. Avis.

Fig. 12. Emys.

Fig. 14.

Pithecus

Fig. 13.

R. O. DEL.

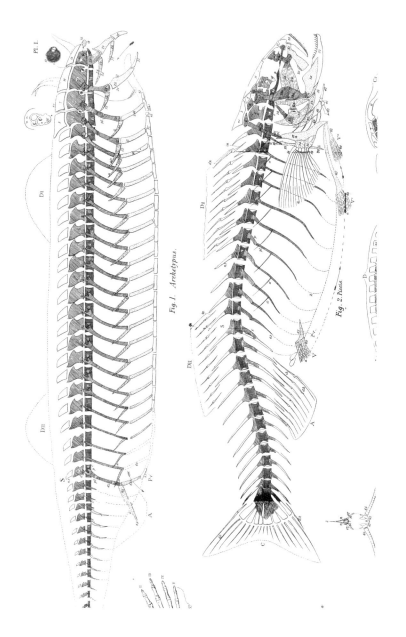

Fig. 1. *Archetypus.*

Fig. 2. *Facies.*

Pl. 1.

Plate 1, detail 3

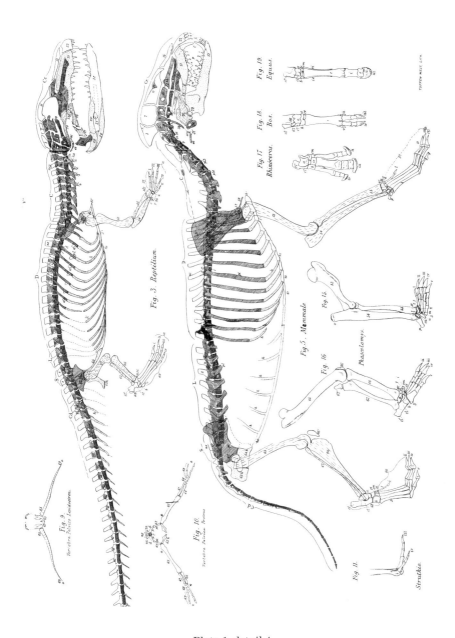

Plate 1, detail 4

Pl. II.

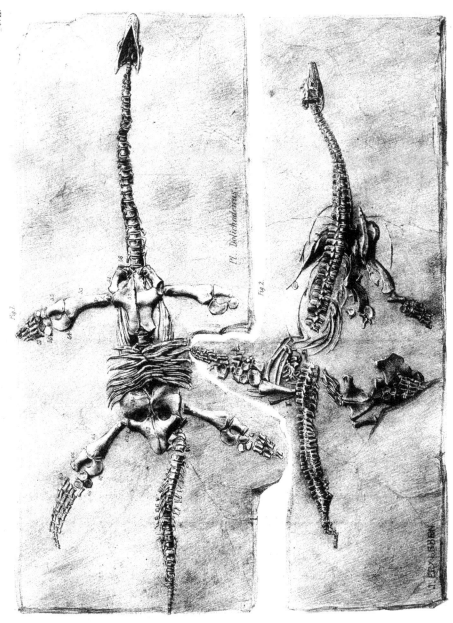

Plesiosaurus Hawkinsii

Plate 2